ROUTLEDGE LIBRARY EDITIONS:
THE ECONOMY OF THE MIDDLE EAST

Volume 26

QATAR
DEVELOPMENT OF AN OIL ECONOMY

QATAR
DEVELOPMENT OF AN OIL
ECONOMY

RAGAEI EL MALLAKH

Routledge
Taylor & Francis Group

LONDON AND NEW YORK

First published in 1979

This edition first published in 2014
by Routledge
2 Park Square, Milton Park, Abingdon, Oxon, OX14 4RN

and by Routledge
711 Third Avenue, New York, NY 10017

Routledge is an imprint of the Taylor & Francis Group, an informa business

© 1979 Ragaei El Mallakh

British Library Cataloguing in Publication Data
A catalogue record for this book is available from the British Library

ISBN: 978-1-138-78710-0 (Set)
eISBN: 978-1-315-74408-7 (Set)
ISBN: 978-1-138-81007-5 (Volume 26)
eISBN: 978-1-315-74451-3 (Volume 26)
Pb ISBN: 978-1-138-82029-6 (Volume 26)

Publisher's Note
The publisher has gone to great lengths to ensure the quality of this reprint but points out that some imperfections in the original copies may be apparent.

Disclaimer
The publisher has made every effort to trace copyright holders and would welcome correspondence from those they have been unable to trace.

QATAR
Development of an
Oil Economy

RAGAEI EL MALLAKH

CROOM HELM LONDON

© 1979 Ragaei El Mallakh
Croom Helm Ltd, 2-10 St John's Road, London SW11

British Library Cataloguing in Publication Data

El Mallakh, Ragaei
    Qatar.
    1. Petroleum industry and trade — Qatar   2. Qatar
    — Economic conditions   3. Economic development
    I. Title
    338.2'7'282095363   HD9577.Q

    ISBN 0—85664—848—5

Printed in Great Britain by offset lithography by
Billing & Sons Ltd, Guildford, London and Worcester

CONTENTS

TABLES AND FIGURES

Tables

This volume is dedicated to my parents-in-law, for their encouragement over the years and because of their deep interest in and concern for the Arab World.

FOREWORD

For almost two decades now I have been involved in studying the Gulf region and its economic development. The absence of an in-depth examination of the economy of Qatar led me to prepare this volume, the goal of which is to present the economic underpinnings of that country's present growth and future development. This small but strategic Arab state juts into the Gulf, about midway between the head of that body of water at Kuwait and Iraq and the Strait of Hormuz, which links it with the Gulf of Oman and the Arabian Sea.

Qatar shares certain characteristics with its Gulf neighbours, primarily the presence of petroleum resources, a small population, the impetus toward rapid economic development, and a capital surplus status. However, the country exhibits special features of its own.

First, education in pre-oil boom days was in advance of many other Gulf states. Numerous leading figures in that region received their training in Qatar.

Second, there is a comparatively strong attachment, by Gulf standards, to the land itself. With a history of trade and mercantile activity and despite an arid climate, Qatar has had a modest agricultural sector. The capital city of Doha, which means 'garden', gives an indication of this attitude.

Third, within the international oil picture, Qatar is not one of the giants. Nonetheless, in OPEC (the Organization of the Petroleum Exporting Countries) it has played its role, sometimes a critical one. The now famous December 1976 OPEC meeting took place in Doha and resulted in a brief period of two-tier pricing. And in the July (Stockholm) and December (Caracas) 1977 and the June (Geneva) 1978 OPEC sessions, Qatar joined what might be called the moderate pricing camp which brought about a price freeze in a crucial period for world energy and monetary stability.

Fourth, Qatar has been a force for stability in the region.

Fifth, the development policy for moderation but not austerity is a refreshing approach to utilising newly acquired affluence.

9

Finally, Qatar has achieved something of a world record in its level of foreign assistance. That aid averages about 12 per cent of the nation's gross national product, the highest proportion of GNP extended internationally by any donor country.

This book seeks to survey the bases and extent of economic development in Qatar and the need to translate the petroleum-generated rapid growth into viable, self-sustaining development. A delicate balancing act is involved, based upon the non-renewable nature of oil and natural gas, that of meeting current requirements while planning for future generations. The chapters offer an introductory overview, trace the development of the oil industry, outline public financing and economic policy, sketch the issues involved in industrialisation, absorptive capacity and agricultural development, examine the social and physical infrastructure as well as money and banking and, lastly, attempt to elucidate the international and regional links in trade, foreign aid, economic cooperation efforts, and investment opportunities.

Officials and planners in Qatar know that they have a finite income or revenue base in petroleum with definable parameters. Thus, domestic, regional, and international investment of present surplus capital funds calls for priority decisions to endow such investment with some of the qualities of an income-generating project. Similarly, domestic spending programmes should result in a self-sustaining level for the Qatari economy when petroleum reserves wane in the coming decades.

The chapters will outline the present direction of government spending priorities and planned projects; future potential areas for domestic investment will be suggested. Private sector investment and spending trends are discussed, based upon the expertise, requirements, and predilections of the non-governmental segment of the economy. Management services should be a major import for Qatar for some time to come.

Hopefully, a number of readers will find this study valuable: teachers and students in such social science disciplines as economics and political science as well as area studies (Middle Eastern); those interested in investment and commerce in that region; readers with an eye to energy and international affairs in a most strategic area of the world today.

My appreciation is due to many individuals who have assisted me in the collection of materials and information. Among them are: Ambassadors Jasem Y. Jamal of the Mission of Qatar to the United Nations, New York, and Abdullah Saleh Al-Mana of the Embassy of Qatar, Washington, DC; Mr Mohamed Noor Al-Obaidly of the Mission of Qatar to the United Nations; the many Cabinet members whom I met and interviewed in Qatar; the Ministers whose offices have patiently supplied data, in particular, the Ministries of Finance and Petroleum and of Information; Mohammed Yousef Mohammed, Director of Power; Ali M. Jaidah, Secretary General of the Organization of the Petroleum Exporting Countries (OPEC), Vienna; Dr Ahmed Abdu with the Ministry of Economy; Abdul Kadir Al-Quadi, Director of Finance; from the Ministry of Finance and Petroleum, Abdulla H. Salatt and Hamed El Ahmady; Kamal Nagy, Ministry of Education; and Kamal A. Saleh, Secretary of the Qatar Chamber of Commerce.

In the preparation of this study I have also had the co-operation and assistance of very capable staff members of the International Research Center for Energy and Economic Development: Dr Jacob Atta, Denise Benzel, Dr Mihssen Kadhim, and Frank Spann. Mrs Glenda Bolin and Mrs Stephanie Dunn faced the task of typing the manuscript which they managed with speed and unfailing good temper. Mrs Cecelia Lange assisted in proofing. Finally, my wife Dr Dorothea El Mallakh served as an unofficial editor; her knowledge of the Gulf and her associate editorship of the *Journal of Energy and Development* gave weight to her suggestions and comments. Any errors or omissions, however, are my own.

The Gulf is a vital and tremendously dynamic area of the world today. Data on the region are still relatively scarce; this volume is meant to bridge, at least in part, that knowledge gap.

Ragaei El Mallakh
July 1978

1 AN OVERVIEW

Introduction

No region of the world has seen so much development activity in the last ten years as the Gulf area. Needless to say, the rapid structural changes which have taken place and will continue to occur in the future in this area are the result of increased oil revenues accruing to the producing countries in the 1970s. A more important aspect has been the greater degree of attention which this oil wealth has focused on the Middle East nations.

Since 'black gold' catapulted the oil-producing countries into the limelight of the international political and economic scene, there has been a proliferation of studies on individual exporting states, especially Saudi Arabia and Iran. However, a majority of the so-called 'small countries' in the oil-producing Arab world have been neglected in this exercise. Despite some articles and short, rather specialised studies, no single comprehensive examination of the economy of Qatar has been published. A number of factors have accounted for this.

First, Qatar is definitely a 'small' member of both the Organization of the Petroleum Exporting Countries (OPEC) and the Organization of the Arab Petroleum Exporting Countries (OAPEC), either in terms of population or by the comparative volume of oil produced annually. Second, Qatar is a relatively new country in the sense that it attained full independence barely seven years ago. Lastly, Qataris are, by nature, modest people: hence, they do not immediately volunteer extensive data on programmes and goals. These factors have given rise to two other complications tending to perpetuate the absence of a comprehensive study on this rapidly developing member of the oil-rich Gulf region. An obvious consequence of this lack of material on official matters in Qatar is the paucity of data and information on the nation necessary for a study of its economic, political, and social system. Second, because of the low-profile information activities of the country, it is not well known in non-Arabic literature and, with 'giants' such as Saudi Arabia and

13

Iran to write about, the labour and time needed for collecting sometimes almost inaccessible material on smaller countries such as Qatar is considerable.

Yet, Qatar has peculiar characteristics which make a study of it valuable. The very lack of information noted earlier coupled with a modest outlook, for example, have their good aspect in that they have created a cautious approach towards development of the country. This approach has spared the nation some of the unpleasant consequences of a rush to industrialise that has plagued other Middle East states. The most recent government budget (1977/8) reflects the commitment to avoid the 'boom-bust' cycle of hyperinflation, in turn followed by overcapacity, then a period of recessionary cooling of the economy — a cycle not at all unfamiliar in the countries of the Gulf. Thus, although the capital budget has increased 25 per cent over the previous year to some QR 3.54 billion ($908 million), only a handful, perhaps 24, of the 200 projects listed are new and those primarily minor in scale. In short, capital spending in 1977/8 will be to continue and/ or complete ongoing projects. Some of the more extensive undertakings have been postponed, including the second airport project, and some scaled down, such as eliminating the university, hospital, and light industrial enterprises.[1]

Another attitude characteristic of the country is that education in Qatar, even prior to the burgeoning oil-generated affluence, was evaluated in advance of many other Gulf states. Numerous leading figures in that region received their training in Qatar. Yet another rather special feature is the comparatively strong attachment, by Gulf standards, to the land itself. With a history of trade and mercantilism and despite an arid climate, the Qataris have had a modest agricultural sector. The capital city of Doha, which means 'garden', gives an indication of this attitude. Finally, while not one of the 'giants', in the world picture Qatar has played its role within OPEC, sometimes a critical one. There was the now famous December 1976 OPEC meeting in Doha which resulted in a brief period of two-tier pricing. (The current Secretary General of OPEC is an Oxford-educated Qatari.) And in 1977 at the July (Stockholm) and December (Caracas) OPEC sessions, Qatar joined what might be called the moderate pricing camp which brought about a price freeze in a crucial period for global energy and monetary stability.

This study will attempt to fill out the perspective on the economic development and rapid growth characterising the Middle East today by offering a description and analysis of the Qatari experience to the growing body of literature dealing with the Gulf region. Working within the data limitations, the following pages strive to delineate the most important economic, social and political forces at play in Qatar and, by extension, within the region.

Background, Government, Geography, and History

During the mid- and late eighteenth century a branch of the Utub tribe in Kuwait known as the Al Khalifa emigrated to Qatar and eventually settled on the northwest coast of the peninsula. The Al Khalifa had a long-established reputation as accomplished sailors and skilled traders, and they soon became the mainstay of commercial development in this region of Qatar.

Further external influences contributed to the early development of Qatar's statehood. By the early nineteenth century the influence of the Islamic sect of Wahabism had spread from the Arabian desert to the Gulf. In fact the British, as the only European power in this region of the world at the time, sought to bring a measure of control to the unstructured, free-moving sea traffic of the Gulf through the colonial government in Bombay. By 1820 the British had persuaded the Arab sheikhs on the then so-called 'pirate coast' to sign a General Treaty of Peace, and in 1835 the First Maritime Truce was signed to outlaw warfare during the all-important pearl-diving season. (Hence, from this 'truce' was derived the term Trucial Coast which applied to the sheikhdoms along the western coastline of the Gulf to the south of Kuwait.) British interest in maintaining stability in the region basically stemmed from a desire to protect the shipping lanes of the Gulf and to the Indian sub-continent since oil had not yet been discovered in the Arabian peninsula.

It was during this time, in the mid-nineteenth century, that the burgeoning power of the Al Thani family was beginning to be felt. The Al Thanis had emigrated from Central Arabia to settle ultimately around Doha. On 12 September 1868 a treaty was signed by Muhammad bin Thani and a British officer, Colonel Pelly, to outlaw war at sea and to refer any conflict to the British Resident of the region. This treaty was similar to other trucial pacts concluded with the British by the neighbouring sheikhdoms.

This period in Qatar's history continued to be somewhat stormy. The Ottoman Turks had since moved into the Gulf region and effectively controlled the port city of Doha. However, the Al Thani family waited to assert themselves once again as the *de facto* rulers of Qatar. In 1879, Muhammad bin Thani's son Sheikh Jasim was appointed Governor of Qatar by the Turks in Doha. In March 1893, Sheikh Jasim seized the opportunity to score a rousing victory over the Turks, although their influence was not totally eradicated. In July 1913, Sheikh Jasim died and was succeeded by his son Abdullah whose crowning achievement was to complete the treaty of November 1916 with the British to become part of the trucial system.

This Treaty of Protection between Great Britain and Qatar lasted until 1 September 1971, when Qatar was declared an independent state under Sheikh Khalifa bin Hamad Al Thani. Although the Treaty of Protection was abrogated, the two states signed a Treaty of Friendship only days later which called for mutual consultation in matters of common concern. This marked the termination of the several other treaty relationships of various forms described earlier which had existed with the United Kingdom since 1896. On 11 September 1971 Qatar was admitted to the League of Arab States. Five days later, the Security Council of the United Nations unanimously recommended Qatar's acceptance into the international organisation followed by ratification on 21 September of that year by the General Assembly.

Qatar had no constitution until the new provisional constitution was announced on 2 April 1970, to become effective that July, making it the first of the Lower Gulf sheikhdoms to adopt such a written document of government. Prior to the announcement all the powers of government were vested in the Emir, the ruler of Qatar.

Article 29 of the provisional constitution declares that the Council of Ministers exists 'to assist the Head of State in the discharge of his duties and the exercise of his powers'.[2] The role of the Council of Ministers is given in Article 37 and 'in its capacity as the highest executive organ of the State, it shall be responsible for the administration of all the internal and external affairs which fall within its competence'. The Council of Ministers may propose and draft laws along with decrees, but must first submit them to the Advisory Council for discussion and for the latter's opinion prior to their submission to the Head of State (the Emir) for ratification and promulgation. Today, there are 15 ministries which represent various aspects of an administration ranked as one of the most efficient in any of the Lower Gulf sheikhdoms.[3]

Another government body is the Advisory Council which consists of four persons popularly elected in each of the ten districts. From these forty elected representatives, the Emir selects two persons from each of the districts, ending with a membership of twenty. The Emir is also entrusted to appoint, at most, three members of the Advisory Council, if he deems it in the public interest to do so. The establishment of the Council represented a significant step toward the gradual transformation of Qatari society from a tribal system of government to a more representative type of socio-political structure. It should be noted that the recommendations of the Council are of an advisory nature, with the Emir empowered to take the ultimate decision on all matters debated by the Council.

Qatar is a small country with a total land area of approximately 4,000 square miles consisting of a peninsula projecting true north into the Gulf for about 100 miles. The entire breadth of the peninsula at its maximum point is only 55 miles. There are also several offshore islands, with only one (Hahlul) inhabited, which contribute to Qatar's total land area.

Qatar's landward frontiers at the neck of the peninsula are shared with Saudi Arabia to the west and Abu Dhabi to the east, while the nearest seaward neighbours are Bahrain and Saudi Arabia. Qatari-Bahraini ties can be explained by a period of dominance over Qatar by the Khalifa family of Bahrain, which ended with the Turkish occupation in 1872. Qatar's ties with Saudi Arabia are based on religion as the ruling families in both countries belong to the Wahabi sect of Islam.

Another close neighbour, Iran, is only 120 miles off Qatar's rounded northern extremity. The nearest Iranian port, Bushire, is 250 miles from Doha, Qatar's administrative and commercial capital situated on the eastern coast of the peninsula. Qatar occupies an important pivotal position on the Gulf for it also lies only 350 miles from the southern Strait of Hormuz, which provides access to the Gulf of Oman and the Arabian Sea. Due to the closeness of ties and/or proximity Qatar has with several Gulf countries, the advantages of regional cooperation are intensified.[4]

The terrain of Qatar is flat except for the Dukhan anticline, which rises from the west coast as a chain of separate hills reaching 325 feet in height. The country is stony, sandy, and barren with natural vegetation consisting of only semi-permanent grazing land restricted to areas around wells, depressions, and short stream sources after winter rain. Precipitation occurs only in winter and early spring and is extremely light, averaging between two and three inches a year. Yet, although not receiving greater rainfall than other nearby states, Qatar is still more fertile than some of its neighbours. The name Qatar may, in fact, have been derived from Arabic for 'a drop of water', or 'qatarah'. The weather is relatively cold in the winter, with invariably great heat during the summer months. In addition to the heat, the humidity is very high around the coastal areas in the summer. The coastal waters are generally shallow, of low tidal range, broken by coral reef.

Not much is known regarding the ancient history of Qatar, although Danish archaeological expeditions between 1956 and 1964 unearthed evidence of prehistoric inhabitation. Excavation of burial mounds and cairns yielded finds of four Stone Age cultures, the oldest almost lower Palaeolithic and the most recent about 6,000–7,000 years old. A complete survey of the Stone Age finds was carried out and the results published by the Danish expeditions.[5] In spite of the pivotal situation of the peninsula, the overall outline of the country did not become recognisable on European maps until well into the nineteenth century.

The Gulf assumed increasing importance as the main sea route to India and the Far East in the latter part of the sixteenth century, although from the dawn of civilisation it was considered the principal trading route between the east and west.[6] Qatar's strategic situation on the Gulf attracted British attention. Interested in safeguarding the sea route to India, throughout the nineteenth century and the early years of the 1900s, Britain pursued a policy of containing the Ottoman presence and influence on the western shores of the Gulf. This policy culminated in the 1916 Treaty noted earlier, and the expulsion of the Turkish garrison from Doha. The treaty is essentially similar to the arrangements concluded with Bahrain, Kuwait, and the Trucial sheikhdoms. Among the features of these treaties which expanded and continued the British influence in the Gulf until the opening of the 1970s was that the Foreign Office in London would handle the 'protectorates'' foreign relations. As a concrete example of the impact of these treaties of protection, one can trace the impact on Kuwaiti oil development. In that case Kuwait could not cede territory to nor have relations with any nation until British consent had been received. The major onshore concession thus involved a joint United States-British venture (British Petroleum and Gulf Oil Corporation as 50-50 partners); the American firm could not be the sole concessionaire.[7]

Prior to the discovery of oil the people of Qatar earned their livelihood by herding in the desert and in the coastal areas by fishing and pearling. Like the other sheikhdoms around the western coast of the Gulf, Qatar was a centre for the pearl industry. The decline of this industry in the thirties, due mainly to Japan's development of cultured pearls, threatened this limited but vital source of income and employment for the majority of the population. The discovery of oil and the beginning of its production brought Qatar the revenues and employment the economy needed at a most opportune time. In December 1949, Qatar became an exporter of crude oil, with its first shipment to Europe. The oil era promised then, and still does, prosperity for Qatar; the impact of this vital material resource on which Qatar's economy and social development hinges is the basic concern of this study.

Population, Labour Force and Productivity

Prior to the exploitation of oil in 1949, Qatar's population was estimated to be between 25,000 and 30,000. The origin of the majority of this population can be traced to two overland migratory movements which occurred in the 1760s and during the period of Wahabi expansion later in the eighteenth century. The migrants in the 1760s came from tribal elements concentrated in Kuwait and along the coast of the Saudi province of al-Hasa. The migrants during the period of Wahabi expansion came primarily from al-Hasa. In both periods, the migrants were bedouins belonging mainly to three famous tribes, the Awamir, Manasir, and Bani Hajir, who were searching for water sources and grazing lands. Other smaller groups of migrants came to Qatar by sea from neighbouring Gulf shores in quest of accessible wells and suitable sites for pearling and fishing.

Today, estimates set Qatar's population around 200,000 with 80 per cent centred round Doha. This high concentration around the capital city can be explained primarily by the large influx of expatriates that began with the development of the oil sector, and to a lesser extent by the internal migration of Qataris from the smaller cities and the countryside into Doha.

The expatriate population is thought to make up approximately half of the country's inhabitants. These non-Qataris come mainly from Pakistan, Iran, Lebanon and Palestine. European contingents are estimated at around 1,500 with the majority being British.[8] The Europeans usually are migrants, planning to stay only a limited amount of time, whereas the others often propose to make Qatar their permanent home.[9]

The influx of vast numbers of foreign personnel, prompted by the inflow of large oil revenues which increased demand for skilled and unskilled labourers, has resulted in repercussions for Qatari authorities. The government, faced with the fast pace of modernisation, has lacked the time to assess and deal with current and future problems of a large expatriate population.

One visible reaction is the shortage of housing which has increased rents and land prices. The development budget of 1977/8 which allocated QR 643 million for housing[10] shows that the government acknowledges the problem and is attempting to alleviate it. The generous public housing programme nonetheless has been unable to meet the rising demand. Consequently, foreign enterprises considering the establishment of offices and expatriate employees in Qatar (and not unlike other Gulf countries) must consider the high cost of living resulting from shortages. David Crawford, the Ambassador for the United Kingdom, has estimated that the cost of establishing a senior executive in Doha is QR 243,000 (or $ 60,900) a year.[11]

The current policy regarding expatriate labour is strict in terms of naturalisation and importation. The labour law of 1962 has not been updated. Article 7 of Decree No. 9 (14 December 1963) states that when applying for a visa, a prospective worker must possess skills beneficial to Qatar and have a Qatari national as a sponsor.[12] A visa, when issued by the Ministry of Labour and Social Affairs, is normally effective for a period of six months and can be renewed indefinitely. Article 1 of Law No. 20 (1963) restricts expatriates from participating in more than 49 per cent ownership of an enterprise, and Article 1 of Law No. 5 (1963) restrains non-nationals from owning any immovable property (i.e., land and buildings) within Qatar. In addition, a foreign worker is not allowed to change jobs if discharged by a sponsor, but is required to leave the country immediately.

This policy of restricting expatriate workers should be viewed from the Qatari government's commitment to the interests of its own nationals in a situation of competition between Qatari citizens and highly skilled and trained foreign workers. The influx of expatriates is premised upon the benefits accruing to those having employment in an economy which is 'booming'. Expatriate labour in Qatar is characterised by three features: (1) the large proportion of earnings remitted to the countries of origin; (2) the fact that many are on loan from their governments or agencies for a specific project or defined time period; and (3) the 1973-4 oil price increases and rising demand for workers in the booming Middle Eastern economies have caused some shortages in various skilled labour specialisations.[13]

The policy of 'freezing' foreign labour to its original occupation tends to increase the need for imported labour, since it limits the possibility of meeting some of the pressure for expanding manpower to higher-productivity employment.[14] Yet the Qatari government has indicated an awareness of such labour problems and is currently attempting to find appropriate solutions. It should be noted that, through the educational structure, Qatar is attempting to alleviate this problem in the longer term by training the indigenous population. The emphasis on human resource development is evident in the new 1978/9 budget with the heaviest outlays in services — housing, education, and health. The Ministry of Education, Culture, and Youth Affairs allocations are up 150 per cent over that of the preceding year.[15]

Despite these labour policy issues, the contribution of non-Qataris to the process of economic and social development in Qatar is substantial. Both in terms of the numerical contribution to the pool of the labour force, especially the skilled, semi-skilled, highly skilled and educated segment, and in terms of the opportunity provided for the indigenous Qatari population to be in constant contact with different outlooks and cultures, the expatriate population has much to offer.

As previously mentioned, present estimates of Qatar's population are around 200,000, including expatriates.[16] Current manpower data are unavailable, with the last official census dated April and May of 1970 (see Tables 1.1, 1.2, and 1.3). Table 1.1 gives evidence of an abnormally high male/female ratio, which can be explained by the high influx of male expatriates. Another result of the expatriate population, as shown in Table 1.2, is the fact that the age groups which constitute the labour force, 15-59, number 66,300 and account for 60 per cent of the total population. Without the non-nationals, the proportion of population within the working age would have been significantly lower at 40 per cent. Due to the present policy regarding naturalisation, the length of time these expatriates will reside in Qatar is unknown, an important consideration in long-range planning. Also shown in Table 1.2 is the high proportion of the indigenous population in the unproductive age groups, which is consistent with other developing countries around the world.[17] As a result of the appreciable oil revenues, Qatar is spared some of the difficulties usually associated with a high dependency load.[18] Nevertheless, this load does require the expansion of the educational, health, and social welfare facilities and service. This development would have direct and indirect repercussions which would be reflected in a continuous demand for both skilled and unskilled labour. In a labour-short and stable economy, the increased demand for workers

Table 1.1: Population of Qatar, 1970

Nationality	Male	Female	Total	%
Qatari	22,668	22,371	45,039	41
Non-Qatari	49,046	17,048	66,094	59
Total	71,714	39,419	111,133	100

Source: State of Qatar Ministry of Economics and Commerce, *The Economic Supply for 1971* (Qatar National Printing, Doha).

Table 1.2: Qatar's Population by Age Groups, 1970

Age in years	Qataris	% of age group	Non-Qataris	% of age group	Total
0–5	10,500[a]	53.8	9,000	46.2	19,500
6–14	13,000	61.8	8,200	38.7	21,200
15–19	3,900	41.9	5,400	58.1	9,300
20–59	14,300	25.1	42,700	74.9	57,000
Over 60	3,300	80.5	800	19.5	4,100
Total	45,000	40.5	66,100	59.5	111,100

[a] Figures are rounded to hundred.

Source: Department of Training and Career Development, *Manpower in Qatar* (Doha, Qatar, March 1974), p. 2.

can only be met if a rising amount of expatriates is allowed to enter Qatar's economy. The import of labour will raise the already high proportion of non-nationals in the total population and thus conflict with the declared governmental policy of Qatarisation aimed at reducing dependency on expatriates. Moreover, the non-Qataris are likely to place extra demands on the existing health and educational facilities in addition to other sectors of the economy. These demands, in turn, will almost invariably be translated into a higher demand for labour.

The distribution of labour in economic activities in Qatar as presented in Table 1.3 indicates a domination by four main sectors in 1970: namely, services, commerce, construction, and government. Foreign labour not only dominates the total labour force, but also all the sectors except the oil industry, where Qataris constitute more than half the total labour force. In 1970, therefore, it would appear that Qataris had left three major forms of economic activity — agriculture, construction, and banking — largely in non-Qatari hands. Especially with respect to agriculture, only 4.2 per cent of its labour requirements were met by Qataris. Thus, the fact that agriculture has been an expanding sector is due in part to the input of foreign labour. Whereas it is not clear whether the low percentage of Qatari labour in the agriculture and construction sectors is due to the often expressed view that Gulf Arabs are only slightly acquainted with such work, it may be hypothesised that the domination of foreigners in financial institutions can be traced to the foreign ownership of many banks and furthermore to the high level of skill demanded in the banking profession, thus inhibiting the development of Qatari banking personnel. But perhaps a more potent reason for the low level of Qatari labour participation in activities other than oil, manufacturing, and government administration is that the aforementioned three activities are considered so important to the economy, hence more Qataris are drafted into them. This may be aimed at ensuring predominance in implementation of the country's development policies regarding oil and industrialisation. Thus, given the already low level of Qatari participation in the total labour force (about 17 per cent), the unavoidable consequence is to leave other sectors such as banking largely to non-Qataris.

The dynamic aspect of industrial distribution of labour in Qatar is brought out clearly by Table 1.4. True to the trend usually associated with the process of development, the agricultural sector has shown the tendency to lose part of its share in the total labour force although, unlike a typical developing economy, the share of agriculture in the total labour force in Qatar has been historically low owing to the arid nature of the land. From the point of view of industrial distribution of labour, therefore, Qatar may not be characterised as a developing country even in 1970. In 1975, both the oil and industrial sectors increased their shares of total employment, the latter by as much as 3.2 per cent while the

Table 1.3: Employment of Labour Force in Industrial Sectors in
Qatar, 1970

Sector	Qataris	%	Non-Qataris	%	Total
Agriculture and fisheries	86	4.2	1,984	95.8	2,070
Manufacturing/quarrying and public utilities	1,825	34.8	3,417	65.2	5,242
Construction	207	2.7	7,578	97.3	7,785
Oil	1,259	57.0	950	43.0	2,209
Wholesale/retail trade	880	11.2	7,005	88.8	7,885
Banking	10	3.3	292	96.7	302
Transport and communications	655	22.0	2,571	78.0	3,226
Government	1,391	22.5	4,781	77.5	6,172
Other services	1,855	13.7	11,644	86.3	13,499
Total	8,168	16.9	40,222	83.1	48,390

Source: Department of Training and Career Development, *Manpower in Qatar*
(Doha, Qatar, March 1974), p. 8.

Table 1.4: Industrial Distribution of Labour in Qatar, 1970 and 1975

Sector	1970		1975	
	Number (1,000s)	%	Number (1,000s)	%
Agriculture[a]	2.1	4.3	3.0	3.0
Petroleum	2.2	4.6	5.0	5.0
Industry[b]	5.2	10.8	14.0	14.0
Construction	7.8	16.1	16.0	15.9
Services[c]	31.1	64.2	62.3	62.1
Total	48.4	100.0	100.3	100.0

[a] Includes fisheries.
[b] Includes quarrying and public utilities for 1970.
[c] Consists of commerce, banking, transport and communications, government and other services, for 1970.

Source: Tables 1.3 and 1.5.

former recorded only a 0.4 per cent increase. It is interesting to note that the total labour force in Qatar, within the five-year period of 1970-5, more than doubled from 48,400 to 100,300, according to Table 1.4.

Another format which may be used to present the structural features of the Qatari economy is to look at the sources of production or the gross domestic product (GDP). Table 1.5 compares the sectoral distribution of employment and GDP for 1975. The comparative analysis of labour and output has implications for labour productivity in the various sectors.

As should be expected, oil dominates the economy by contributing 72 per cent of the GDP, but employs only 5 per cent of the total labour force in Qatar. Of course, the oil industry is notoriously capital intensive. Nonetheless, these figures imply a high level of labour productivity in this sector. On the other hand, services, which employ 62.1 per cent of the total labour force, contribute 16.7 per cent of GDP, which shows that the services sector is not as productive as other sectors in terms of the labour-output ratio. The last column of Table 1.5 reveals the labour productivity differential between oil and other sectors of the Qatari economy even more vividly. What is surprising, however, is the relatively high labour productivity in the agricultural sector, compared, for example, to the services productivity. There are, nevertheless, very good reasons for this spectacular achievement of agriculture. For one thing, agriculture in Qatar is highly capital intensive to the extent that substantial capital expenditures have been undertaken to provide supporting services, including irrigation, research, and the like, to this sector. On the other hand, the services sector is traditionally highly labour intensive. Further, the fact that this sector includes the government, which does not depend entirely on economic criteria to measure productivity, should account for the low output-labour ratio in this sector. As noted earlier, the high labour productivity in the petroleum sector is due primarily to the capital-intensive nature of the industry.

After all, Qatar is a progressive economy in the Arab world. Its development has been remarkably rapid despite labour and other factors which constitute constraints on the country's development efforts. The future development of the

Table 1.5: Sectoral Distribution of Employment and GDP for Qatar, 1975

	Estimated employment		Estimated Gross Domestic Product[a]		
	Number (1,000s)	%	Value (million riyals)	%	Output-labour ratio (1,000 riyals)
Agriculture	3.0	3.0	60.8	0.8	20.3
Petroleum	5.0	5.0	5,250.0	71.9	1,050.0
Industry	14.0	14.0	311.7	4.3	22.3
Construction	16.0	15.9	458.5	6.3	28.7
Services	62.3	62.1	1,220.8	16.7	19.8
Total	100.3	100.0	7,301.8	100.0	72.8

[a] The GDP figures were estimated using 1965 Kuwaiti labour productivity estimates.

Source: US Department of State, *Medium Term Ability of Oil Producing Countries to Absorb Real Goods and Services* (CACI, Inc., Arlington, Virginia, March 1976), pp. 13-18.

country depends, in large measure, on the ability of its government to strike a workable balance between the economic need for foreign labour and the social disamenities which often accompany a foreign-dominated labour market. This makes it abundantly clear that Qatar should strive for an overall approach to the population issue. That a governmental response in this direction is under way can be seen in the 1977/8 budget allocation emphasis.

General Framework of the Study

Granted that manpower is perhaps the greatest single constraint on development, what can the country do to ensure a fair degree of economic development? The answer to this is found in the gradualistic approach which the government of Qatar has adopted towards economic and social development. As the discussions in both the chapters on industrialisation, absorptive capacity, and agricultural development and on public finance and economic policy will indicate, the government's attempt to develop has been based on a moderate tone. However, the former chapter also underscores the problems which both physical and manpower constraints have created and will continue to create for the efforts of the country to absorb the increasing oil revenues. One way to raise the absorptive capacity of the nation has been to expand social and welfare services and physical infrastructure. The activities in these areas are covered in yet another chapter. Investments in these sectors not only act as avenues for current spending but also serve to increase future absorption, for example, by providing a more efficient labour force.

However, continual expansion of absorptive capacity presupposes that the resource base of the country will continually expand. To the extent that at least in the near future the economy of Qatar will continue to depend on oil for its development, the second chapter on oil traces the role which the oil industry has played in Qatari development and will continue to play in the future. This has implications for the future resource or revenue base of Qatar. The section of this study centring on industrialisation, absorptive capacity, and agricultural development (as has been noted) also yields an evaluation of what has been done and will be effected in the future in the area of diversification. One objective of the diversification and balanced growth policy is to ensure an adequate resource base for future development when oil no longer constitutes an important source of revenue and foreign exchange.

The overall trend of government finance (revenue and expenditure) is contained in the chapter on public finance and economic policy. With respect to policy, it would appear that fiscal and monetary policies cannot be separated in the Qatari case, as drawn from the analysis in the segment devoted to money and banking. In addition to economic policy, however, the money and banking chapter also examines the financial and banking system in Qatar, especially the recent developments which have been brought about by the oil boom. If the economy of Qatar were to be described in any other way apart from its oil base, that description would be that it is liberal and very open. Qatar depends almost entirely on international trade for its livelihood. International trade is treated as a separate unit to show the extent of this dependence. That chapter also considers the important aspect of Qatari development within a regional context and its foreign aid activities not only to the immediate area but to Africa, other developing countries, and international agencies. Finally, a brief survey of investment opportunities in Qatar is offered, a subject of no little interest to the oil-importing nations with balance-of-trade and payments problems.

Notes

1. Chase World Information Corporation, *Mideast Markets*, 10 April 1978, p. 2. (Hereafter cited as *Mideast Markets*.) In that interview, the technical advisor to the Emir noted: 'We have an Emir who has always favored slow, deliberate growth. . . . A couple of years ago, for instance, we were given a study saying we needed 25 to 30 new berths at Doha port. Now . . . we have eight berths there and a couple of those are standing empty.'

2. This and other citations from the constitution are from the Ministry of Information, Government of the State of Qatar, *Qatar into the Seventies* (Ali bin Ali Printing Press, Doha, 1973), pp. 11-12. (Hereafter cited as *Qatar into the Seventies.*)

3. The Ministries include: Foreign Affairs; Economy and Commerce; Justice; Defence; Water and Electricity; Interior; Industry and Agriculture; Municipal Affairs; Finance and Petroleum; Public Works; Labour and Social Affairs; Transport and Communications; Public Health; Information; Education, Youth, and Culture. John Duke Anthony, *Arab States of the Lower Gulf: People, Politics, Petroleum* (The Middle East Institute, Washington, DC, 1975), p. 75.

4. The chapter on international trade discusses regional cooperation in the Gulf area in which Qatar could participate.

5. It should also be recalled that archaeologists ranging up and down the Arab states of the Gulf have uncovered significant relics of a once thriving major centre of civilisation hitherto overlooked. This is described in Geoffrey Bibby, *Looking for Dilmun* (Alfred A. Knopf, New York, 1969). Brief descriptions of the work on Qatar are given in chapter 7, pp. 117-44 and pp. 165-8.

6. For an interesting account of the Gulf's history since the sixteenth century see Riad N. El-Rayyes, *The Oasis Rivalry and Oil: The Problem of the Arabian Gulf, Between 1968-1972* (El Harar, Beirut, 1973). In Arabic.

7. Ragaei El Mallakh, *Economic Development and Regional Cooperation: Kuwait* (University of Chicago Press, Chicago, 1968), pp. 10-11.

8. The Permanent Mission of the State of Qatar to the United Nations, *The Era of Reform* (New York, 1973), p. 32.

9. Special report, *Middle East Economic Digest* (hereafter MEED), April 1977, p. 23.

10. *Middle East Economic Survey* (hereafter MEES), 24 January 1977, p. 12. The monetary unit is the Qatari riyal (QR) with an approximate exchange rate of 4QR = US $1.

11. MEED, 9 September 1977, p. 31.

12. Diplomats, international visiting businessmen, and their families are exempt from these requirements.

13. Other Gulf states face similar labour problems, e.g. Kuwait in Ragaei El Mallakh, *Economic Development and Regional Cooperation: Kuwait*, p. 123.

14. United Nations Industrial Development Organization, *An Industrial Survey of Qatar*, UNIDO/TCD 103, April 1972, pp. 47-9.

15. *Mideast Markets*, 10 April 1978, p. 2.

16. The 1970 official census estimated a population of 111,133; in 1974 according to the International Monetary Fund, *Survey*, 19 August 1974, p. 259, population was set at 180,000; and MEED, special report, April 1977, p. 3, stated the current figure of 200,000. (Hereafter International Monetary Fund, *Survey*, will be cited as IMF *Survey*.)

17. For a comparison of this ratio with those of developed and under-developed regions, and the implications thereof, see Hans W. Singer, *International Development: Growth and Change* (McGraw-Hill, New York, 1964), pp. 80-5.

18. For a discussion of the implication of high dependency ratio, see Joseph Spengels, 'The Population Obstacle to Economic Betterment', *American Economic Review*, Papers and Proceedings, May 1951, pp. 343-54.

2 DEVELOPMENT OF THE OIL INDUSTRY

Introduction

The oil sector in Qatar is currently the only sector capable of generating a sizeable amount of foreign exchange. In 1975 some 71.9 per cent of the country's gross domestic product (GDP) was generated directly by the oil sector, and government receipts from this sector for the same year constituted 92.8 per cent of total revenue. Through a policy of diversification it is hoped that dependence upon petroleum will be reduced in the future. It has been estimated that, by 1985, Qatar could conceivably reduce the oil sector's percentage contribution to total GDP to about 58 per cent.[1] This will be achieved mainly through expansion of services, with only a slight increase in agriculture and a more significant growth in industry. Nevertheless, no matter how future sectoral distribution of Qatar's GDP is projected, it is agreed that the oil sector clearly will dominate the economy in the near and medium term.

With oil playing such a predominant role as a source of revenue, the development goals of Qatar hinge upon the policies affecting this sector. Such issues as pricing, level of production, and integration of the oil sector to other sectors of the economy have serious consequences for the current and future development of Qatar.

Structure and History of the Oil Industry

Before the discovery of oil, Qatar's pearling industry was the mainstay of the economy. During the 1930s Japan began producing and marketing cultured pearls, an action which had disastrous effects on Qatar's earning potential. Thus, Qatar has already experienced the difficulties which can result from overdependency upon a single resource. The discovery of oil was welcome, indeed, for it came immediately after the slump in the pearl market. Now the Qatari government aims at diversification as one means of maximising the benefits from petroleum.

Another policy the government has recently pursued is to

maximise the gains from the petroleum sector through acquisition of 100 per cent ownership and control over all areas of oil production. The government began participation in oil operations in 1973 when it took over 25 per cent ownership of both the Qatar Petroleum Company (QPC) and Shell Qatar. With a participation agreement, even a 100 per cent agreement, the government is given the option of partaking in various operations such as exploration, sales, and production at the level of its choosing or of leaving these responsibilities to the concessionary company. The change, of course, lies in the fact that the original company is not the sole owner, if an owner at all, of the oil produced. If the company retains a share in the oil produced, the crude belonging to it is referred to as 'equity crude'; the oil owned by the government is the 'participation crude'. In addition to taking an equity share of the petroleum, if shares are owned, the company usually 'buys back' a percentage or the total amount of the government's share. The quantity of oil involved in the buy-back arrangement is determined by the ability of the government to sell the oil to other customers at a price which is higher than the buy-back price. Then the host governments are left to receive payments in the form of taxes and royalties on the companies' equity oil and from the buy-back price on oil received for their own share.

In the past, the Gulf countries, including Qatar, have not had much success in selling their oil at a price higher than the buy-back level. The concessionary companies have been in a position where the price of their own equity oil has amounted to only two-thirds the buy-back price. As a result, they have been able to offer petroleum at a price somewhere between the buy-back and equity costs.

To counter the leverage given to the concessionary companies in the pricing structure, Qatar instituted a token reduction in posted prices and raised, by a much higher amount, tax and royalty rates applied to the companies' equity oil. The resulting effect was that Qatar could offer oil at a lower price than the concessionary companies. This causes the companies either to reduce their profit margins and/or to increase the market share of oil sales in Qatar's favour.

The full participation agreements which the Qatari government recently obtained are an outcome of 40 years of co-operation with foreign oil companies. In 1937 Petroleum Developments (Qatar) Limited, an operating company of Iraq Petroleum Company Limited, began oil exploration in Qatar encouraged by the discovery of oil in Bahrain five years earlier. The results were successful and Qatar's discovery well, Dukhan Number 1, located at the northern end of Western Jebels began production. Soon afterwards two additional wells were drilled on this same site. With the outbreak of World War II, production ended and was not resumed until late in 1947. It was then that the first shipment of Qatari crude bound for Europe was made from the newly completed terminal at Umm Said.

Petroleum Developments, changing its name in June 1963 to Qatar Petroleum Company Limited (QPC), was owned by British Petroleum (23.75 per cent), Royal Dutch/Shell (23.75 per cent), Compagnie Francaise des Petroles (23.75 per cent), Standard Oil of New Jersey now Exxon (11.875 per cent), Mobil Oil (11.875 per cent), and Participations and Exploration Corporation (the Gulbenkian family estate, 5 per cent). The QPC was responsible for onshore production at the Dukhan wells.

As a result of the improvement of offshore drilling and exploration techniques, two United States companies, Superior Oil Company and the Central Mining and Investment Corporation, were granted offshore exploration concessions in 1949. The area covered under the agreement included all seabed areas beyond the three-mile territorial limit which fell under Qatar's jurisdiction. These companies soon forfeited their concessionary offshore rights when exploration proved unsuccessful. Subsequently, Shell Company-Qatar (SCQ), owned by Royal Dutch/Shell,[2] resumed exploration with rights to all of Qatar's territorial waters excluding the Qatari islands. In 1976 this area was reduced to 5,000 square kilometres. Three offshore fields resulted, with production beginning at Idd al-Shargi in 1964, Maydam Mahzam in 1965, and at Bul Hanine in 1972. The Idd al-Shargi field was the first seabed field in the world to be operated completely as offshore facilities.[3] Today these three fields are connected to Halul island in the north, a storage and pumping centre.

In addition to these offshore fields and the Dukhan wells

there is the Bunduq field, shared equally with Abu Dhabi. The field is managed by Abu Dhabi Marine Areas (ADMA) with Qatar's concession going to the Bunduq Company, owned by British Petroleum, Compagnie Française des Petroles, and a group of Japanese oil companies called United Petroleum Development Company of Japan, with each holding equivalent shares.

In 1973, two years after Qatar's independence, the government gained control of 25 per cent ownership of both QPC and SCQ. At that time, QPC and SCQ also agreed to reduce their share to 49 per cent by 1980. The agreement was short lived and in February 1974 the Qatari government signed and ratified agreements with QPC and SCQ which provided for 60 per cent state participation in the rights and operations of those companies. The new 60-40 agreement was modelled on the arrangement between the government of Kuwait and British Petroleum/Gulf reached a month earlier. In addition, the agreement was the first of its kind between an assemblage of major oil companies and a Gulf state specifying buy-back prices. The Gulf producing states had previously held that 93 per cent of the posted price should be buy-back price for participation crude.[4] This was the price the Qatari government was able to agree upon with both the onshore producer, QPC, and the offshore producer, SCQ.

In December 1975, the government of Qatar announced that negotiations would begin for full control of all oil production. Less than a year later, in September 1976, the remaining 40 per cent controlled by QPC was turned over to the government. This agreement differed from the Kuwaiti model, which Qatar had originally proposed, in that it was more along the lines of QPC's terms. The main difference was that QPC would receive a service fee for each barrel of oil produced, not only for liftings as in Kuwait. This fee was to serve as remuneration for the management and operational services which QPC would continue to supply. In addition, this service fee is indexed directly to the government's official selling price for Dukhan crude. Thus, the fee will vary in direct proportion to the price changes of the onshore Dukhan crude. At the time of the agreement, the fee was 15 cents per barrel of onshore production, with the price per barrel at $11.85. The fee paid would remain constant as the same percentage of the selling price (1.266 per cent). This means that at current prices (as of 1 January 1978) the companies receive approximately 16.7 cents per barrel (see Table 2.1). The companies will also receive £18 million (around $31 million at exchange rates for that time) 'in consideration for its [i.e., the government's] acquisition of the remaining QPC in the rights, assets, operations, and facilities in Qatar'.[5] This amount was determined to be slightly over book value. Furthermore, the agreement reduced the amount of Dukhan crude the companies would be able to lift at the official government sale price from 150,000-170,000 barrels a day (b/d) to 130,000 b/d for the period of five years, with the additional understanding that the volume of liftings cannot vary by more than 15 per cent in either direction.

The agreement reached with SCQ for 100 per cent participation in offshore oil production was not signed until February 1977. The terms are very similar to those affecting onshore production. The service fee was set at 15 cents per barrel of oil produced from the existing offshore fields (Idd al-Shargi, Maydam Mahzam, and Bul Hanine) and subject to change in the official selling price for Qatar Marine crude, the crude oil produced from the offshore fields. This fee will also apply to natural gas liquids (NGL) from the projected NGL-2 plant scheduled for completion in 1979. For compensation SCQ received £14 million (around $24 million) for the remaining 40 per cent interest in offshore production which it held. In addition, SCQ has agreed to purchase at the official government sale price 145,000 b/d of Qatar Marine crude oil for the next five years, with an understanding, as with QPC, not to vary the lifting by more than 15 per cent.

As a result of the changes in Qatar's oil industry, reorganisation in the government was necessary. The Ministry of Finance and Petroleum has been entrusted with making policy decisions, with the Department of Petroleum Affairs accountable for the development of the oil industry in Qatar. A national oil company, Qatar General Petroleum Corporation (QGPC) was established to monitor the government share in all oil, gas, and petrochemical operations in foreign countries and within Qatar.

Various operations (see Table 2.2) fall under the jurisdiction of the QGPC and include the Qatar Petroleum Producing

Table 2.1: Posted Prices of Qatar Crude Oil (US $ per barrel)

Date of posting	QPC (Umm Said)	SCQ (Halul)
from the start to		
7/15/53	1.830	—
7/16/53	2.080	—
5/28/57	2.210	—
2/13/59	2.030	—
8/16/60	1.930	—
2/13/64	—	1.830
2/15/71	2.280	2.200
6/1/71	2.387	2.305
1/20/72	2.590	2.501
1/1/73	2.705	2.614
4/1/73	2.862	2.766
6/1/73	3.025	2.923
7/1/73	3.804	2.980
8/1/73	3.200	3.092
10/1/73	3.143	3.037
10/16/73	5.834	5.503
11/1/73	5.899	5.564
12/1/73	5.737	5.412
1/1/74	12.414	12.013
11/1/74	12.014	11.613
10/1/75	12.740	12.540
12/31/76	11.850	11.660
1/1/77	13.190	13.000
7/1/77	13.190	13.000
1/1/78	13.190	13.000

Sources: Various issues of the *Petroleum Economist* and *Petroleum Intelligence Weekly*.

Authority (QPPA) as one of the most significant. The QPPA is divided into two sections: first, the QPPA Onshore, which is essentially the old QPC, and second, the QPPA Offshore, which handles the former SCQ concession. The QPPA received a budget of $ 220 million in 1977 to cover investment and operating costs. This amount excludes the 1977 allocation of $ 70 million for the second NGL plant and $ 6 million

Table 2.2: QGPC Shareholdings

Company	Capital	QGPC share (%)
Abroad		
Arab Maritime Petroleum Transport Company	$ 500 m	13.75
Arab Petroleum Pipelines Company (SUMED)	$ 400 m	5.00
Arab Shipbuilding & Repair Yard Company	$ 100 m	16.00
Northern Petrochemicals Company (COPINOR)	100 m Fr. fr.	40.00
Qatar		
National Oil Distribution Company	$ 40 m	100.00
Qatar Fertiliser Company		
preference shares	QR 42,428,500	100.00
ordinary shares	QR 50 m	70.00
Qatar Petrochemicals Company	QR 240 m	80.00
Qatar Gas Company	QR 400 m	70.00
QPPA Onshore		100.00
QPPA Offshore		100.00

Source: *The Oil Industry in Qatar*, 1976.

for a seismic survey.[6] The operations of the QPPA are managed by Dukhan Service Company for onshore activities and by the Qatar Shell Service Company for offshore operations.

Another important operation under the direction of the QGPC is the National Oil Distribution Company (NODCO), which owns and operates the refinery at Umm Said in addition to regulating the sale of all oil products subsidised by the government.

Thus, Qatar has achieved full operation and control of the oil industry. Along with this has come the financial responsibility for further exploration and research accompanying day-to-day operations. At the same time, Qatar is able to benefit from the oil companies' expertise in management and technical abilities.

Production and Reserves

It has been estimated that Qatar's oil resources will last for 30 years before they are depleted.[7] With output averaging close to 450,000 b/d, Qatar is the eleventh largest producer in the Organization of Petroleum Exporting Countries (OPEC); Qatar ranks fifteenth in world output. On a production per capita basis, Qatar is third, just behind Abu Dhabi and Kuwait. In addition to oil reserves, Qatar has large quantities of associated gas (gas produced with oil) and has recently discovered unassociated gas (gas found not in connection with oil). With the completion in 1975 of the first NGL plant, Qatar began to rely more heavily on the utilisation of its gas reserves with oil expected to play a somewhat reduced role in the future.

Oil

Production of oil in Qatar began in 1949, marked by the first export of oil to Europe. This shipment originated from the onshore Dukhan field which today has 94 wells.[8] The Dukhan field yields crude of 41.5° API, which is of high quality, having an average sulphur content of 1.1 per cent. There are three main centres in the field: Khatiya, commissioned in 1949; Fahahil, commissioned in 1954; and Galeha, commissioned in 1955.

In 1964 the SCQ began putting the offshore areas into production with the Idd al-Shargi field, which is capable of producing 15,000 b/d. The following year Maydam Mahzam was brought into production and, at capacity, yields 155,000 b/d; the most productive field yet brought on stream is the Bul Hanine, which has an ouput capacity of 160,000 b/d. The latter field has been estimated to possess the greatest amount of offshore reserves, a little under 1 billion barrels at current prices and today's level of technology. These offshore holdings yield a crude of lower density than those produced from Dukhan, with 36° API and average sulphur content of 1.52 per cent.[9] As a result, onshore crude has historically obtained a higher posted price (see Table 2.1).

Owing to the fact that the production of oil in Qatar has approached its peak, Sheikh Abdel-Aziz Bin-Khalifa, Minister of Finance and Petroleum, announced in early 1976 that oil production would be held around 475,000 b/d.[10] A current estimate of average daily production shows that Qatar Marine crude (offshore) and Qatar Dukhan (onshore) supply 250,000 b/d and 225,000 b/d, respectively.[11] It therefore

Map 1: Situation Map, Qatar Concessions

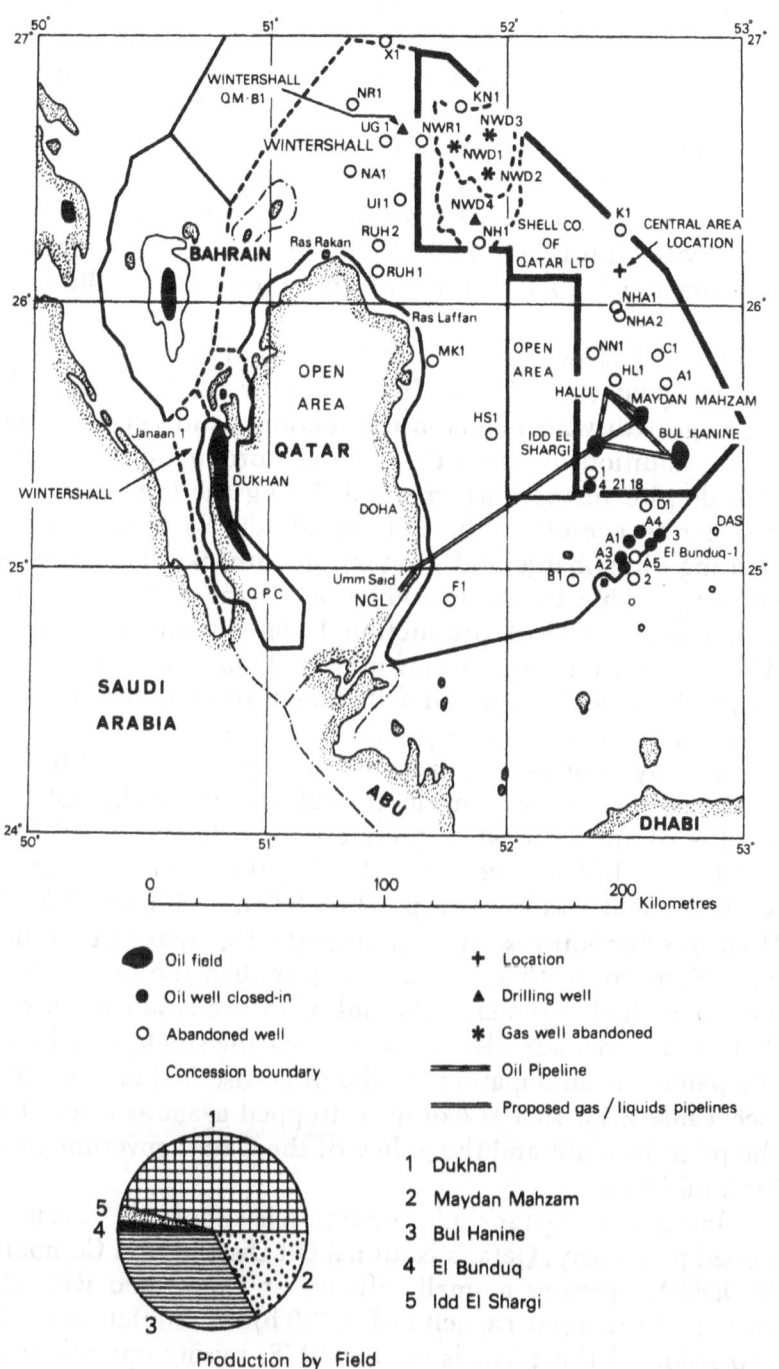

Production by Field

1 Dukhan
2 Maydan Mahzam
3 Bul Hanine
4 El Bunduq
5 Idd El Shargi

appears that Qatar has been able to maintain production at a level concurrent with policy decisions. Output capacities for the onshore fields have been calculated at 350,000 b/d and for the offshore at 300,000 b/d. Although production, of course, reflects world demand for Qatar crude oil, the government chooses not to exceed the target of 475,000 b/d in an effort to extend oil receipts.[12] Excluded from this target figure is the production from Bunduq oil field, which Qatar shares with Abu Dhabi. In 1976, the first year of production, an average of 20,407 b/d was produced at the Bunduq field.[13]

Table 2.3 shows the historical pattern of crude oil production in Qatar since 1949. The oil sector has had relatively steady growth with jumps in production, as in 1964, reflecting the addition of new wells. In 1973 production reached a peak despite restrictions imposed during the last quarter of the year as a result of the October Middle East war and the resulting oil embargo and production cutbacks. The increase was largely due to the introduction of the SCQ's new Bul Hanine field. In 1974, production declined when Qatar, in an effort to assure longer earning potential, imposed a ceiling on oil production. Production fell even further in 1975 when competing crude oils were priced lower than Qatari crude oil. It was only possible to reach the average of 437,600 b/d in 1975 due to a massive production effort during the last few months of the year when production reached a record high of 610,000 b/d in December. During the previous summer, when demand was low, output had fallen to 300,000 b/d. In 1976, production rose again despite the fact that Qatar crude was selling more than 50 cents higher than the price would have been had it retained the link with the Arabian market. Part of this increase also resulted from the stockpiling by oil companies in anticipation of the price rise in January 1977 (see Table 2.1). In 1977 output dropped again as a result of the price increase and the policy of the Qatari government to limit oil liftings.

Almost all of Qatar's oil production is exported.[14] As mentioned previously, Qatar's National Oil Distribution Company (NODCO) operates a small refinery at Umm Said with the current throughput capacity of 6,000 b/d. Completion of the expansion of this plant is set for 1978, raising operations to

Table 2.3: Crude Oil Production in Qatar since Commencement
(in thousand barrels)

Year	Daily average	Total	Cumulative	Annual (%) change in daily production
1949	2.0	730	730	
1950	33.6	12,268	12,998	1,580.0
1951	49.3	18,009	31,007	46.7
1952	69.2	25,255	56,262	40.4
1953	85.0	31,025	87,287	22.8
1954	99.9	36,450	127,737	17.5
1955	115.0	41,983	165,720	15.1
1956	124.2	45,345	211,065	8.0
1957	138.5	50,558	261,623	11.5
1958	175.1	63,910	325,533	26.4
1959	170.4	62,197	387,730	− 2.7
1960	174.6	63,908	451,638	2.5
1961	177.2	64,675	516,313	1.5
1962	186.2	67,980	584,293	5.1
1963	191.5	69,884	654,177	2.8
1964	215.3	78,813	732,990	12.4
1965	232.6	84,902	817,891	8.0
1966	291.3	106,307	924,198	25.2
1967	323.6	118,100	1,042,298	11.1
1968	339.5	124,266	1,166,564	4.9
1969	355.5	129,746	1,296,310	4.7
1970	362.4	132,261	1,428,571	1.9
1971	360.7	157,206	1,585,777	18.8
1972	482.4	176,543	1,762,320	12.0
1973	570.3	208,160	1,970,480	18.2
1974	518.4	189,216	2,159,696	− 9.1
1975	437.6	159,724	2,319,420	− 15.6
1976	479.0	178,120	2,497,540	9.6
1977	432.0	160,820	2,658,360	− 9.8

Sources: Organization of the Petroleum Exporting Countries (OPEC), *Annual Statistical Bulletin, 1975*; *OPEC Bulletin*, 20 February 1978; and *Petroleum Economist*, March 1978.

Table 2.4: Qatar's Crude Oil Exports, 1970-5 (in thousand barrels per day)

Destination	1970	1971	1972	1973	1974	1975
North America	–	*1.7*	*3.5*	*10.9*	*77.9*	*88.1*
of which:						
USA	–	1.7	3.5	10.9	77.9	88.1
Latin America	*9.6*	*10.2*	*4.8*	*0.7*	–	*14.8*
of which:						
Argentina	8.2	–	–	–	–	–
Brazil	1.4	10.2	4.8	–	–	–
Western Europe	240.4	*294.1*	*335.7*	*360.8*	*291.2*	*235.0*
of which:						
France	46.0	59.1	45.0	69.9	79.5	32.4
Germany (FR)	–	5.2	5.3	2.2	14.2	20.4
Italy	46.0	72.3	57.7	65.8	57.0	42.9
Netherlands	31.3	50.3	122.5	144.4	16.3	75.0
Sweden	22.2	25.3	15.0	8.0	14.2	1.4
Turkey	–	–	4.4	1.8	–	–
United Kingdom	93.5	76.4	72.9	64.8	97.0	54.8
Middle East	*1.9*	*4.3*	*7.8*	*9.0*	*4.9*	*1.7*
of which:						
Southern Yemen	1.9	4.3	7.8	9.0	4.9	1.7
Africa	*49.9*	*56.6*	*41.1*	*89.9*	*27.3*	*19.8*
of which:						
South Africa	41.0	44.4	28.5	61.9	–	–
Asia and Far East	*36.2*	*36.8*	*55.8*	*73.5*	*99.0*	*41.9*
of which:						
Japan	3.5	1.5	1.1	3.7	4.7	1.4
Philippines	–	–	8.3	3.6	8.4	–
Thailand	32.7	35.3	44.4	57.6	49.3	27.9
Oceania	*24.8*	*14.9*	*17.2*	*11.5*	*10.7*	*4.9*
of which:						
Australia	24.8	13.5	11.7	11.5	10.7	4.9
Unspecified[a]	–	*10.1*	*15.8*	*14.0*	*0.2*	*22.1*
TOTAL	362.8	428.7	481.7	570.3	511.2	428.3

[a]   To Gibraltar for wireless order.

Source: Organization of the Petroleum Exporting Countries (OPEC), *Annual Statistical Bulletin, 1975*, p. 70.

50,000 b/d and allowing for export of refined products in addition to satisfying domestic requirements.

The United States has been the single largest importer of petroleum from Qatar (see Table 2.4). The volume of exports to the United States jumped from 1.9 per cent to 15.2 per cent of total oil exports in 1973 and 1974, respectively. No doubt the 1973 figure would have been larger had it not been for the oil embargo imposed with the outbreak of the Middle East war in October of that year.

Natural Gas

Although Qatar has reached a peak in oil production and can see an eventual end to its oil reserves, the recent discovery of the North West Dome unassociated natural gas, which may prove to be one of the largest fields in the world, reduces the uncertainty for Qatar's future. Until the completion in 1975 of the first natural gas liquids plant, NGL-1, associated gas was generally flared off. Domestic industries take advantage of the residual gas, with the NGL broken down into propane, butane, and gasoline for export. Ethane in the future will serve as a feedstock in the petrochemical plant currently under construction. Associated gas from the offshore fields is presently flared. With the completion by the end of 1979 of the second natural gas liquids facility known as NGL-2, this too will be utilised. As with the onshore associated gas reserves, the different natural gas liquids (propane, butane, and natural gasoline) will go to expand exports, the ethane into feedstock for petrochemicals, with the difference being the future for the residual gas, which at the moment is uncertain.

The unassociated gas, found in very large quantities, was discovered by Shell 40 miles northeast of the tip of the peninsula. It is estimated at about 100 trillion scf in this new find known as the North West Dome (NWD). The discovery lies in the Permian Khuff strata some 10,000 feet below the surface; four wells have now been drilled into this structure. The real speculative excitement results from the possibility of future unassociated gas fields still to be revealed. A geological feature referred to as the 'Qatar arch' turns north-south through the entire peninsula and at the tip changes to a northeast direction where it ends at the offshore boundary.

The NWD reservoir is located at the top end of the Qatar arch which indicates the possibility of further gas reserves throughout all of Qatar.

Unassociated gas also exists under the existing Dukhan and offshore fields. Thus far, exploration of the offshore unassociated gas fields has disclosed only small quantities of reserves. The Dukhan field has proved more promising; nonetheless, an accurate estimate of reserves is not possible at this time.

The prospects of unassociated gas in Qatar are currently being investigated through an exploratory seismic survey undertaken by Prakla, a German company. If the results of this survey prove advantageous, the Qatar government will presumably lease out onshore tracts with profit-sharing agreements. Other companies have been involved in offshore Qatari exploration including Wintershell (a German firm, which is still operating its concession), the Japanese-owned Qatar Oil Company, and Holcar Oil Company; the latter two firms have relinquished their concessions.

Two problems seem to emerge from Qatar's future reliance on natural gas. First, the amount of worldwide reserves of natural gas available is uncertain. The long-term future market for natural gas and its by-products appears fruitful, yet discoveries in the North Sea, Siberia, Alaska and the Canadian Arctic could place those sources closer to potential industrialised consumers. The result would be that the Qatari natural gas could find it difficult to compete in the marketplace at a future date. Second, Qatar received a serious setback in April 1977, when its Umm Said NGL plant (using Dukhan field gas) was destroyed through an explosion and fire. This facility was an important link with gas and other projects in Qatar. Nonetheless, rebuilding is under way as well as the establishment of new facilities to expand gas utilisation.

Determinants of Revenues from Oil

Oil revenue which accrues to the Qatari government is so important to the strength of the national economy that factors which determine its level or flow ought to be analysed in a cohesive manner. The government's oil-related revenue is influenced by the following elements: (a) production level;

(b) selling price; (c) structure of royalty, fee or tax rates affecting the oil industry; and lastly, (d) the extent of the Qatar government's participation in the oil industry. These four factors over the years have determined the flow of oil revenue to the government; the extent of influence of each has fluctuated. However, there has been a general positive and upward trend in the 1970s (with the exception of production level) in the changes which have taken place in price structure, fees, profit tax and royalty payments, and, above all, state participation. Therefore since the cost of crude oil extraction to the government can be considered as an insignificant proportion of the total value of oil, government revenue from this source has shown marked acceleration.

If Qatar were to handle its entire oil production, without, for example, either intermediaries or management contractors, then one could represent the total revenues accruing to Qatar as the product of the quality of oil extracted and the price at which each unit of oil (i.e., barrel) is sold: oil revenue = (quantity of barrels of oil) X (price per barrel). Consequently, the rate of growth of oil revenue will be equivalent to the sum of the rates of growth of oil production and the price at which each barrel is sold. For the purpose of initial analysis, let us assume that oil revenue is determined solely by quantity and price. Then Table 2.5 gives an idea of the potential that the Qatar government might have had in the 1960s given different central patterns than those which existed. This also underlines why the trend was toward full government participation in the oil economy. With 1960 as the base year, oil production rose from an index of barely 50 in 1953 to over 200 in 1970. The rate of increase in extraction was smooth, except for the 1958 to 1960 span when the production level stayed almost constant or declined (1959). The rapid oil exploitation continued from 1970 until it reached an all-time peak of 208.2 thousand barrels (570,300 b/d) in 1973, or 325.7 per cent more than the level in 1960; lifting began to slow in response to conservation measures introduced by the Qatari government. Unlike production, the price of Qatari oil until 1971, when OPEC's strong bargaining power began to tell on the international oil scene, showed a declining trend. The index of 107.8 in

Table 2.5: Index of Qatar's Annual Oil Production and Price[a]
(1960 = 100.0)

Year	Production	Price	Year	Production	Price
1953	48.5	107.8	1966	166.3	100.0
1954	57.0	107.8	1967	184.8	100.0
1955	65.7	107.8	1968	194.4	100.0
1956	71.0	107.8	1969	203.0	100.0
1957	79.1	114.5	1970	207.0	100.0
1958	100.0	114.5	1971	246.0	120.9[b]
1959	97.3	105.2	1972	276.2	134.2
1960	100.0	100.0	1973	325.7	222.9[b]
1961	101.2	100.0	1974	296.1[b]	632.9[b]
1962	106.4	100.0	1975	249.9	660.1
1963	109.4	100.0	1976	279.7	614.0
1964	123.3	100.0	1977	251.6	683.4
1965	132.9	100.0			

[a]   Posted price is used as a proxy for selling price of a barrel of oil.
[b]   These are years in which several price revisions took place. The index uses a simple arithmetic average.

Source: Computed from Tables 2.1 and 2.2 above.

1953 (which held until 1957) rose to 114.5 in the two years 1957 and 1958; the unilateral company-announced price cuts in 1959 and 1960 brought the index down to 100, where it remained unchanged for 11 years until 1971.

It will be recalled that it was the 1959 and 1960 slashes in posted prices which were largely responsible for the creation of the Organization of Petroleum Exporting Countries as a producer-nation response.

Consequently, it can be said that prior to the 1970s oil production in Qatar followed an inverse supply curve: the lower the price, the greater the quantity which was produced. This situation is obviously contrary to economic theory for it would suggest that oil companies operating in Qatar, and indeed in other Gulf petroleum-rich countries, did not follow a profit-maximising motive by cutting down on production when prices fell. Yet they did follow a profit motive, a version which cannot be explained by the negative correlation between production and price, for positive-sloping supply applies only if the producer is distinct from the buyer. In the case of multinational oil companies, however, in most cases the producer and buyer were the same company. Even where there is a separate buyer, the actual price at which a barrel of Qatari oil was sold was higher than the posted price in Qatar. The posted price therefore might have been kept at a low level for the purpose of royalty and tax payments computations. The Qatar government did not benefit from the higher actual sale price.

A pricing turnaround began with the 1970s after OPEC had time to acquire experience and stronger bargaining power. The details of the role of OPEC will be examined later. One conclusion can be offered at this stage: despite the sluggishness in the behaviour of oil price, the almost consistent increases in production indicate that theoretically if Qatar had been in complete control of its oil industry in the 50s and 60s, its revenue would have experienced a consistent upward swing. However, practically speaking, one has to consider the limitations of the country at that time in terms of technical resources and manpower to operating oil development on the scale which actually obtained. Additionally, the oil companies have had an essential role not only in production but also in transport and marketing.

The non-availability of data for oil revenues in the period prior to 1970 puts a constraint on our ability to make any quantitative judgement on actual growth of Qatar's governmental revenues as contrasted with the potential growth evidenced by the rapid increase of production of oil. We, however, with the aid of Table 2.6, carry out a comparative analysis of growth of oil revenues which may be attributable solely to the rates of growth of output and price on the one hand, and the actual growth rate as experienced between 1971 and 1977. It is from this table that some quantitative assessments of the contributions of oil production, price, and the other two factors (extent of government participation and royalty/tax/fee structures) may be made. The argument is as follows: since column 5 is the sum of the rates of growth of production and price (i.e., columns 2 and 3), it follows that if the actual rate of growth of oil revenues (column 4) is greater than column 5, then factors other than

Table 2.6: Rates of Growth of Qatar's Oil Production, Posted Price and Government Oil Revenues, 1971-7

Year	Production	Price	Actual oil revenues	Potential oil revenues
(1)	(2)	(3)	(4)	(5) = (2) + (3)
1971	18.8	20.9	63.1	39.7
1972	12.3	11.0	30.2	23.3
1973	17.9	65.9	39.8	83.8
1974	− 9.1	184.2	336.4	175.1
1975	− 15.6	4.3	− 6.1	− 11.3
1976	11.5	− 7.0	24.7[a]	4.5
1977	− 9.7	11.3	− 1.5[b]	1.6

[a] Provisional.
[b] Estimated.

Source: Computed from Table 2.5 above, except column 4 figures which were derived from statistics in the subsequent chapter on public finance.

those affecting oil price and output must have contributed to the difference, enabling the government to obtain more revenue from oil over and above those attributable to an increase in production and a price rise. On the other hand, if column 4 is less than column 5, as was the case, for example, in 1973, then it follows that Qatar could not reap full benefits from the export value increase. In the rest of the analysis, an attempt will be made to explain the divergence between the growth rate of oil revenues and the potential rate of growth in columns 4 and 5, respectively.

Since direct participation in the oil industry of any kind by the government did not take place until 1973 (when the government took over 25 per cent ownership of QPC and Shell Qatar), the substantial growth in actual oil revenues in 1971 may be credited to either increased royalty rate, fees, or tax rate of oil companies' profits, or a combination of all three. It is conceivable that some of the revenue which should have been paid in 1970 did not actually flow into the government's coffers until 1971. However, a more direct reason was the promulgation of Law No. 21 of 1970 which raised the profit tax rate of oil companies from 50 to 55 per cent. Also, the new concession agreement reached with the Qatar Oil Co. Ltd (Japan) might have brought in some of the QR 54.3 million bonus payments promised at the signing of the agreement, apart from the QR 10.9 million initial signature-bonus paid and the QR 483 thousand annual concession rental promised.

Since the state took over 25 per cent of the interests of the oil companies in January 1973, one would expect that the actual rate of growth of revenues would exceed the combined rates of production and price for that year. Table 2.6 shows this did not happen. A feasible explanation is that there might have been a one-year lag between the accruing of the revenue and the actual payment of all of it to the government. Thus, when in 1974 a new agreement enabled the government to acquire 35 per cent more of the oil companies' assets, actual revenues increased at a rate almost double that consistent with the combined rate of growth of output and price. Indeed, some of the 1974 oil revenues due spilled over into 1975 payments, thus offsetting some of the negative effects of a 15.6 per cent reduction in oil production (despite the fact that oil price went up by 4.3 per cent in that year).

The future of oil revenue for the government of Qatar is bright after the 100 per cent participation agreement but for the uncertainties in the international monetary market, with special reference to the dollar in which unit the Qatari crude oil price is quoted. This is not a problem which is peculiar only to Qatar. A discussion of the effects of the volatility of the dollar's value in the world market on the pricing policy of OPEC countries will be offered in the forthcoming section concentrating on OPEC.

Qatar and Multilateral Oil Organisations

A study of the petroleum industry in Qatar, or indeed of the economy of any oil-producing country in the Gulf region, will be incomplete without an examination of the role which OPEC has played in production and pricing policies internationally.

Qatar has been active in the Organization of Petroleum Exporting Countries (OPEC) and the Organization of Arab Petroleum Exporting Countries (OAPEC) since their respective inceptions. In fact, the December 1976 meeting of the

OPEC Ministers was hosted by Qatar, and its sessions focused attention on the events occurring in Doha. It should be remembered that it was this meeting which temporarily divided the OPEC nations over the level of the increase in crude oil for the first half of 1977. Although some speculated that the Doha Conference would mark the start of OPEC's disintegration or gradual weakening in its power, a reconciliation was soon reached at the mid-year meeting; the organisation continues as a strong force in crude oil pricing and hence production policies. It was over the issue of crude oil pricing that OPEC was founded in 1960. At that time, the producing countries' revenues were determined by the posted price of crude oil under the 50-50 profit-sharing arrangements. The posted price, which differed from the market price, was reduced unilaterally by the international oil companies thus causing a drastic decline in the revenue for the producing states. The result of the February 1959 slash in the posted price led to the creation of OPEC a few months later after a second reduction. Since then, the OPEC nations have concentrated on a variety of issues of mutual concern. As a member of OPEC, Qatar was involved in the various transformations of the oil industry, such as changes in royalty payments, pricing, taxation, and finally, the 100 per cent participation agreements between the producing countries and the international oil companies.

From the behaviour of posted prices in the 1960s, it would appear that the formation of OPEC did not result in any increases in oil prices until 1971. Although OPEC was not completely ineffective during the first 11 years of its existence, its real strength did not emerge until the opening of this decade. The test of its effectiveness in relation to crude oil price ought to be a comparison between the constant price of crude oil, which prevailed from 1960 to 1971, and the price which would have obtained had there been no OPEC. It is believed that the formation of the organisation may have prevented further reductions by the oil companies of posted prices below the 1960 level in response to low prices prevailing in the consuming nations during this period.

The oil organisation came of age, so to speak, in the 1970s when negotiations with the oil companies produced positive results in both areas of oil price and state participation. OPEC made it possible for member countries to negotiate as a unified body (sometimes at regional levels) for both higher prices and greater involvement in the production of crude oil.

The first two years of OPEC's operations were concentrated in studying the nature of the world oil industry. In 1962, the body recommended that its members negotiate with the oil companies in their respective countries for prices equal to those obtained in 1959. This accounts for the freezing of the Qatari posted price at $1.93 per barrel between 1959 and 1971. In addition, a comprehensive study of price structures of the oil market was launched with a view to relating the posted prices of oil to the prices of OPEC imports. In other words, since the ultimate use of foreign exchange obtained from oil revenues is to pay for imports, the OPEC states sought to ensure at least a constant purchasing power for a unit of oil revenue.

At the beginning of the current decade, the oil market tilted in favour of oil-producing countries, as a result of the emergence of excess demand. This encouraged further negotiations between the producing nations and the producing firms. Thus, the Caracas (Venezuela) meeting in December 1970 provided a forum for the passing of several measures, including the adoption of uniform posted price increases by member countries. Also, members were urged to establish a 55 per cent minimum income tax rate on oil companies, and allowances, which heretofore had been deducted from profit before tax, were to be eliminated as of 1 January 1971.[15] This was to be incorporated into the Tehran Agreement in February 1971.

The now famous Tehran Agreement was the cumulative result of a series of events which began with the pace-setting measure won by Libya in its negotiations with oil companies operating in the country. Indeed, the 'Libyan scenario'[16] might have been the motivating factor underlying the Caracas objectives. Another Libyan step at the negotiating table in January 1970, barely four months after the first agreement, together with pressures from Venezuela, forced the companies to agree to enter into new negotiations with OPEC members. Thus the Tehran meeting was convened, and the

Tehran Agreement was born.

Among other things, the agreement provided that the profit tax levied on oil companies be raised from 50 to 55 per cent. Also a uniform 33 per cent per barrel price increase was to be effected immediately; in addition, further price adjustments were to become effective at later dates, some of which were meant to serve as insurance against inflated imports. The elimination of allowances hitherto deducted by oil companies before income tax, together with the other provisions of the agreement, all tended to assure the oil-producing countries of increased revenues and stable prices at that time. However, when the dollar was devalued in December 1971, it was clear that the Tehran Agreement had become outdated, at least so far as its provision on prices was concerned. A re-negotiation was called for by the countries involved, yielding a formula for pricing which hopefully would insulate the oil-generated revenue from exchange rate movements. This is the pricing scheme which would tie the price of oil to the so-called Geneva I basket of international currencies.

With the strength of OPEC successfully tested, the organisation forged ahead to negotiate for increased participation of member states in their respective oil industries. At the July 1971 conference, a resolution was passed to enhance participation. From the viewpoint of Qatar, the relevant development towards participation was a series of negotiations which Minister Ahmed Zaki Yamani of Saudi Arabia conducted on behalf of the Gulf countries with the oil companies. After some setbacks, an agreement was finally reached between the Petroleum Minister and the oil companies operating in the Gulf region on 5 October 1972. The general participation agreement formed the basis for negotiations between the governments of the producing countries in the Gulf and the oil firms operating in each nation.

Among other things, the agreement provided for an initial 25 per cent participation, to rise to 30 per cent in January 1978, and thereafter to increase by 5 percentage points annually until 1981, after which a 6 percentage point increment would bring the total participation as of 1 January 1982 to 51 per cent. Further, the agreement contained provisions which, if implemented, would ensure a continuous market for the government's share of crude output resulting from the participation. Although many of the provisions of the participation agreement were set aside for the benefit of oil-producing countries, it was a cornerstone in the long process toward achieving national control over a vital source of economic development by the Gulf petroleum states.

These agreements in the areas of pricing formulae and participation reflected both the increasing power of the producers within their organisation of OPEC through coordinated action and the tilting of the scales toward a 'seller's' market in petroleum. The 1973-4 OPEC-set price increases marked a watershed in this respect. Despite temporary gluts in the world oil market (as in 1977-8), generally speaking the advantage will lie with the producing countries throughout the remainder of this century. Moreover, the high quality of Qatar's oil should serve as an additional edge to that nation's bargaining position in international petroleum trade.

Another aspect of OPEC which is often overlooked is the international role it plays not directly related to crude oil. As a bloc, it has pledged aid to developing countries and concerns itself with the total global economic order.[17] Qatar's assistance through OPEC will be discussed in the final chapter on international trade and regional cooperation.

Contrary to some Western opinion that OPEC would not survive for long after the 1973 oil crisis, there is no doubt in the minds of OPEC officials and heads of member countries that the organisation has come to stay for some time. The instability in the world monetary market and the need to have a common front in marketing problems not only for crude, but for the products of downstream operations, will tend to fortify OPEC and make it an integral part of the international oil industry.

In contrast, OAPEC (Organization of the Arab Petroleum Exporting Countries) was founded under broader objectives related to regional needs of certain of the oil-producing nations.[18] These aims have resulted in joint projects such as the Arab Petroleum Investment Corporation, the Arab Maritime Petroleum Transport Company, and the Arab Shipbuilding and Repair Yard Company. In this manner through OAPEC Qatar has participated in regional projects, information exchange, and regional marketing efforts to best utilise its petroleum industry.

In conclusion, the Qatari economy rests heavily upon the policies related to the oil sector. Since independence in 1971, this sector has been subject to several changes, especially in pricing. The transitions, such as the gradual 100 per cent participation and mutually negotiated agreement, have followed general policies of the other OPEC nations. Moreover, the Qatari government does not hesitate to initiate various strategies on its own related to oil, as evidenced by the limit on production set at 475,000 b/d.

Qatar has attempted to extend the revenues from the oil sector, as oil is a depleting natural resource. The nation is well aware of this limit as estimates as to the life of its reserves do not appear to stretch very far into the future. Fortunately, the discovery of vast quantities of natural gas will prolong revenues earned directly from the petroleum sector.

Notes

1. Farid Abolfathi *et al*, *The OPEC Market to 1985* (D.C. Heath & Co, Lexington, Massachusetts, 1977), p. 235. This estimate is a 'best forecast' which makes various assumptions regarding future government policy which the Qatari government may or may not follow.

2. In 1967, SCQ made an agreement with Ente Nazionale Idrocaruburi (ENI) which gave the Italian company rights to 20 per cent of overall production under a long-term purchase agreement in return for investment in exploration and production.

3. *Pakistan Monitor*, 22 February 1973, p. 8. All three fields bear the names of long-standing pearling beds.

4. MEES, 19 April 1974, p. 2.

5. MEES, 27 September 1976, p. 16.

6. *Financial Times*, 9 March 1977.

7. Organization of the Petroleum Exporting Countries, *OPEC Bulletin*, 21 November 1977, p. 6.

8. Sixty of the wells produce oil, four gas, five control and monitor water pressure, 21 produce water, and four are closed. The water is used to raise the pressure and extend the life of the well.

9. This is an average figure derived from total crude exported from the storage and pumping terminal on Halul island.

10. MEED, 2 April 1976, p. 23.

11. *Petroleum Intelligence Weekly*, 13 March 1978, supplement.

12. *Financial Times*, 9 March 1977.

13. MEED, 22 July 1977, p. 26.

14. See the chapter on international trade and regional cooperation for a detailed account of Qatar's export market.

15. A.A. Kubbah, *OPEC: Past and Present* (Petro-Economic Research Centre, Vienna, Austria, 1974), p. 23.

16. Ibid.

17. For a discussion on the role of OPEC and the world economic order, see Nazar Al-Khalaf, 'OPEC Members and the New International Economic Order', *The Journal of Energy and Development*, Spring 1977, pp. 239-56.

18. Refer to George Tomeh, 'OAPEC: Its Growing Role in Arab and World Affairs', *The Journal of Energy and Development*, Autumn 1977, pp. 26-36.

3 PUBLIC FINANCE AND ECONOMIC POLICY

The Role of the Government in Qatari Economic Development

Introduction

Of all the surplus-fund countries of the Arabian Gulf, Qatar has shown the greatest awareness of the fact that oil is an exhaustible resource.[1] In recognition of this, the government in Qatar holds that oil revenues ought to be utilised efficiently through a prudent expenditure policy, making certain that sufficient surpluses are kept for future development. Nonetheless, governmental spending in this country has been undertaken, so far, without a general direction towards a comprehensive development plan. Whether the absence of a development plan has reduced the effectiveness of the government's budgetary operations (or their efficiency) is by no means a question which can be answered easily.

A conjecture may be made however that, since the country is very small (both in terms of land area and population size),[2] and furthermore, given the high population concentration in the Doha area, perhaps a comprehensive development plan is a superfluous economic institution. Consequently, it may be argued that the necessary central direction which a comprehensive development plan is expected to give to budgetary allocations could be provided, in such a small and compact economy, by the office of the Emir, under the supervision of the Director of Finance. The conception, initiation, and supervision of development programmes appear to be a shared or cooperative approach rather than centralised in any single institution utilising a formal procedure or process. Indeed, until the ascension of Sheikh Khalifa Bin Hamad Al Thani to the throne in 1972, there was no clear-cut development policy in the country.[3]

Objectives of the Qatari Development Policy

As mentioned earlier, Qatar's development policy at the moment is not contained in a single comprehensive economic plan. In 1972 a British firm of consultants was requested to

56

launch an all-encompassing survey of the development requirements for the entire country. The study emphasised seven areas where economic and social development should be stressed.[4]

1. Housing. A popular housing scheme with public assistance and methods designed to promote the private sector has been adopted with the goal of increasing the number of houses available to 17,000 by the 1990s.
2. Education. Twenty-four elementary and 14 secondary schools are scheduled for completion in 1982.
3. Health. Expansion of existing facilities, a 650-bed hospital in Doha and ten health centres are to be completed by the 1980s.
4. Development of Doha. The plan includes an international airport and reconstruction of Doha with two industrial centres.
5. Diversification of the economy. An attempt is being made to broaden the production base so as to reduce the dependence on crude oil exports. The projects include a steel plant, and a petrochemical complex, along with fertiliser and cement plants.
6. Revival of traditional sectors. The encouragement of agriculture and the fishing industry by government assistance and the provision of modern facilities are expected.
7. Import substitution. Goods can be produced for domestic consumption such as plastic and leather goods, soap, edible oils and paper.

Toward the end of 1973 the government adopted the plan. Thus, government expenditures and projects place more emphasis on long-term development. The goals of the development policy should also include: the attaining of a better income distribution through education, training, and the expansion of economic opportunities in general; developing the abilities of the indigenous population so that they may be able to participate fully in and accelerate the process of economic and social development; enlarging the absorptive capacity of the economy so that more capital can be employed productively as well as channelling surplus capital into secure and productive investment regionally and internationally; and finally, coordinating Qatar's industrialisation programme with the overall development of the Arabian Gulf region.[5]

The difficulty in obtaining a clear picture of the development objectives and programmes of Qatar is due in part to the less structured and more diversified management of the Qatari economy, especially in matters relating to finance and surpluses. The discussion of public finance is important because, in lieu of a development plan, it is only in the analysis of budgetary estimates and allocations that the role of the government and its developmental objectives could be elucidated. What are the roles which the government in Qatar is expected to play in the country's economic development? Based upon the enunciated policy that the government has to respond to the immediate and longer-term welfare of both the present and future generations of Qataris, the set of roles which the government ought to play should necessarily include those which would ensure improved opportunities for the present generation and assure an adequate standard of living for the country in the future. In this sphere, the government's role is virtually unlimited, to the extent that all the oil revenues accrue in the first instance to the government. Clearly, the present and future welfare of the population is intricately tied to the manner in which the government spends the accumulated oil revenue.

Government Revenue and Expenditure Analysis

The data available indicate that, since 1970 and at least until 1976, Qatar consistently has enjoyed budget surpluses. The reasons for these surpluses may be broken down into two categories: (a) those due to increased revenues, and (b) factors retarding the growth of expenditures to the extent that expenditures are unable to grow at a rate fast enough to absorb the accumulated oil revenues.

The elements which have influenced the growth of revenues are essentially those which have increased oil revenue in general, especially in the 1970s. Oil revenue constitutes more than 90 per cent of total revenues of the government.[6] This component of government income has increased with rising petroleum output, increased government participation in the oil industry, and, above all, high posted prices for crude, particularly since the energy crisis in 1973. With respect to expenditures, the proverbial constraints on absorptive capacity in surplus countries, namely manpower, management, and physical bottlenecks, are to share the blame for the relatively slow growth of expenditures. However, in the case of Qatar, these factors may be less responsible than in other Arabian Gulf states for the moderate rate of growth of government expenditure, especially prior to 1974. A more important cause is the conscious policy of the Qatari government not to rush into vast development programmes and overheat the economy. The authorities recognise that efficient initiation and implementation of large programmes require technical manpower and management, which are in short supply in the country, and furthermore demand considerable experience on the part of the government in order to monitor them; hence the gradualistic approach towards development in Qatar.[7]

Government Finance

Postponing for the moment the discussion of the composition of revenue and expenditure, we may look at the movement of these fiscal variables over time as presented in Table 3.1. It will be seen that even before the marked jumps in oil prices in 1973 and after, government revenues in Qatar had shown a consistent tendency to exceed government expenditure. Indeed, between 1970 and 1973, government revenue increased at an average annual rate of 44.4 per cent while the rate for expenditure was lower by 5.5 percentage points. However, owing to the leap in revenue in 1974 (325.5 per cent over 1973 figure), the period 1974-7 experienced a relatively faster growth of revenue at an annual rate of 87 per cent, while expenditure grew by 54 per cent. Over the entire seven-year period (1971-7), revenue grew faster than expenditure by an average of 20.7 percentage points per annum, compared with 33 percentage points in the last four years of the period.

The description of the rates of growth of revenue and expenditure explains the pace at which the government of Qatar has been adding to its budget surpluses. The average annual growth rates of this residual element in the government finance have been 94.4, 283.2, and 202.3 per cent

Table 3.1: Qatar: Government Finance for Fiscal Years 1970-7[a]

(QR million)	1970	1971	1972	1973	1974	1975	1976[b]	Budget 1977
Revenue	579.4	945.1	1,230.4	1,720.0	7,319.0	7,135.0	8,811.0	8,948.0
Expenditures	504.6	689.8	958.6	1,352.0	2,464.0	4,432.0	5,894.0	7,319.0
Surplus or deficit (—)[c]	74.8	255.3	271.8	368.0	4,855.0	2,703.0	2,917.0	1,629.0
Growth rates (percentages)								
Revenue		63.1	30.2	39.8	325.5	−2.5	23.5	1.6
Expenditures		36.7	39.0	41.0	82.2	79.9	33.0	24.2
Surplus or deficit (—)[c]		241.3	6.5	35.4	1,219.3	−44.3	1.9	−44.2

a The Qatari fiscal year is identical with the Islamic calendar (Hijra) year which is 11-12 days shorter than the Gregorian calendar year. The fiscal years 1970-7 are the Hijra years 1390-97 which end respectively, on the Gregorian calendar days 25 February 1971; 14 February 1972; 4 February 1973; 24 January 1974; 12 January 1975; 2 January 1976; 21 December 1976; and 10 December 1977.

b Provisional.

c Includes net lending and equity participation.

Sources: International Monetary Fund (IMF), *International Financial Statistics* (IMF, Washington, DC, September 1977), p. 300; IMF *Survey* (15 August 1977), p. 259.

for 1971-3, 1974-7, and 1971-7, respectively.

On the whole, therefore, despite the fact that spending has risen rapidly in Qatar in the 1970s, revenue has recorded an even faster growth to maintain the high rate at which surplus was accumulated by the government. Several qualifications, however, ought to be noted at this juncture: 1974 was definitely an exceptional year for government finance in Qatar; the jump in revenue was astronomical, and as a result, although expenditure recorded the highest annual rate of growth in this year, it was insufficient to prevent the surplus from spiraling upward by 1,219.3 per cent. The non-uniformity of the growth in revenue is glaringly brought out by the rates for the four years 1974-7. There was a decline in revenue in 1975, resulting in a reduction of surplus by as much as 44.3 per cent in that year. Further, the budget estimates indicate that 1977 was a repeat of the 1975 situation. It would appear, therefore, that the average annual rates used in the discussion could give the false impression that all was bright in Qatari government finance so far as budget surpluses are concerned in the 1971-7 period.

The second qualification which should be made to this analysis is that surplus includes net lending and equity participation by the government of Qatar. The data which are available to us separate these expenditure items from 'surplus proper' for only the 1973-7 financial years. Therefore, to obtain a consistent set of data for the 1971-7 period, this separation was ignored in Table 3.1. Thus, for the 1974-7 financial years, if net lending and payments for equity participation by the Qatari government in business ventures are treated as expenditures, the growth of surplus becomes drastically less than the rates shown in the table. For comparison, the rates of growth of surplus after netting out equity participation payments and net lending by the government, are: 1,984.3, −75.7, −16.5, and −367.5 per cent for the 1974, 1975, 1976, and 1977 financial years, respectively.[8]

Revenue Composition

Oil dominates the sources of government income in Qatar to such a degree that one may be tempted to ignore the analysis of the revenue structure; nonetheless, to understand the extent of dependence on the oil sector earnings such an exercise is instructive. Table 3.2 gives the necessary information base for analysis.

A significant aspect evident from the table is that, despite the increases in revenue originating from oil after 1973, revenue derived from non-oil sources has continued to show a steady share of total revenue. It was only in 1974 that, in spite of a doubling of revenue from non-oil sources (from QR 104 million in 1973 to QR 266 million in 1974), the share dropped to an all-time low of 3.6 per cent. In the years which followed, non-oil revenue was able to recapture its position by increasing at a rate faster than that of oil revenue. The 1977 budget expected non-oil revenue to constitute about 9 per cent of the total.

Although the tax laws relating to profit-incomes exempt companies which are in 'joint ventures with the government, under contract to the government or sub-contracting to the government'[9] from the payment of income tax, the non-oil revenue of the government has increased significantly in the recent past. This is primarily due to the buoyancy of the economy in recent years, especially in the private sector. Also, with rising imports and the recent upward revision of the duty on cement, increased revenues from import duties have emerged, although a large number of imported items attract very low duties.

The dominance of oil in Qatari government revenue is likely to continue in the years to come, unless serious efforts are made to diversify the composition and sources of revenue. An issue which may be raised in the revenue analysis is whether the absence of purchase, sales, and labour income taxes as well as the exemptions on imports and company profits should not be questioned in light of the need to find alternative governmental revenue sources when the capacity to export huge quantities of crude oil is gone. So far as labour income is concerned, the no-tax policy might be considered an incentive for foreign workers to seek employment in Qatar. Furthermore, from the point of view of Qataris, it is one way to ensure a high standard of living. The exemption of some imports from duties and the low rates applicable to most of them also may be justified on the same grounds that any increase in import duties will fan inflation by providing a higher base from which domestic prices

Table 3.2: Qatar: Composition of Government Finance, 1973-7[a]

	1973	1974	1975	1976	Budget 1977
Revenue					
Total	1,720	7,319	7,135	8,811	8,948
Oil[b]	1,616	7,053	6,623	8,262	8,138
	(94.0)	(96.4)	(92.8)	(93.8)	(90.9)
Other[b]	104	266	511	549	810
	(6.0)	(3.6)	(7.2)	(6.2)	(9.1)
Expenditure					
Total	1,352	2,464	4,432	5,894	7,319
Foreign grants[c]	370	522	892	306	431
	(27.4)	(21.2)	(20.1)	(5.2)	(5.9)
Other[c]	982	1,942	3,540	5,588	6,888
	(72.6)	(78.8)	(79.9)	(94.8)	(94.1)
Net lending and equity participation[b]	183	999	1,767	2,134	3,721
	(10.6)	(13.6)	(24.8)	(24.2)	(41.6)
Surplus or deficit (−)[b]	185	3,856	936	782	−2,092
	(10.8)	(52.7)	(13.1)	(8.9)	(−23.4)

[a] See note a, Table 3.1, for the relationship between the Qatari fiscal and Gregorian calendar years.
[b] Parentheses are percentage of total revenue.
[c] Parentheses are percentage of total expenditure.

Sources: As for Table 3.1.

will rise. Business profit tax exemptions could have the objective of making the Qatari market more attractive, particularly to foreigners with an eye to government contracts. Since the oil boom in the Middle East has generated accelerated requests for tenders in all the oil-rich countries, competition for top foreign firms can develop.

Nevertheless, there are adverse facets to all these measures.

First and foremost, the country cannot depend on crude oil exports for 90 per cent of its revenue indefinitely; there is a commitment to go on generating enough revenue to support development at a rate consistent with the high standard of living envisaged. Even with the diversification policy of the government, it may be necessary after the oil is effectively finished to expand the tax base of the economy. If this is a future possibility, then it is necessary, at least over a long period of time, to train the Qatari population to the need for taxation in order that the government can provide social and other services. To ensure that the standard of living of the majority of the population is not adversely affected during this 'learning process', the amount of taxes collected could be re-injected into the economy through increased subsidies. The proportion of the taxes which is pumped back into the system would depend of course on the strength of other sources of revenue. Indeed, it is felt that a real distribution of income in Qatar, with a view to raising the standard of living of the majority of the population, cannot be achieved quickly without a mechanism of a progressive system of taxation. This implies that the tax system to be implemented in the 'learning years' should necessarily exempt a large majority of income earners (i.e., low-income earners) from paying tax.

Taxing imports also would correct the imbalance which might exist between the prices of locally produced and of imported goods. However, this is not in reality a pressing problem for Qatar at the moment to the extent that most of the imports do not have locally produced substitutes. As the industrialisation programmes of the country are diversified to produce import substitutes, however, the need to protect the local industries will become obvious. That the Qatari authorities are aware of this possibility is exemplified by the cement industry case. Import duty on cement was raised from 2.5 per cent to 40 per cent to protect the local output of this product.[10] One also may argue that perhaps a careful examination of the composition of imports might reveal there is room to employ import duties to cut down some ostentatious and luxury consumption imports which have been generated by the oil wealth. Prudence in the use of foreign exchange is necessary if the economy is to reap optimum benefit from oil revenues simply because that revenue is derived from a wasting asset.

Expenditure Pattern

In the analysis of a country's government expenditure, one is interested in several elements. What is the breakdown into current and project (or capital) expenditures? What is the composition of capital expenditure and the implications on policy shifts which may be drawn from changes in the relative shares of capital expenditure items? Specific to the Qatar case is the question whether one can examine the capital expenditure structure and make inferences regarding the seriousness the government attaches to the diversification objective in the development process. Data do not exist for a sufficient number of years on development expenditure composition. There are two government expenditure structures. The first is that contained in the lower half of Table 3.2, which is readily available for 1973-7. This expenditure pattern shows how surplus or deficit and net lending and equity participation by the government, both in absolute amounts and as percentage of total government revenue, have been changing between 1973 and 1977. Also in that structure, current and capital expenditures are not distinguished. One could look at the growth of foreign aid, vis-à-vis 'other' expenditure, essentially current and capital domestic expenditures combined, less payments for equity participation and net lending by the government. Thus one of the features of the government expenditure pattern shown in Table 3.2 is that while there have been fluctuations in absolute terms, since 1973, there has been a falling trend in the share of Qatari expenditures earmarked for foreign aid as indicated by a drop from 27.4 per cent of the total in 1973 to an estimated 5.9 per cent in 1977; this can be compared to an increase in 'other' expenditures from 72.6 per cent to 94.1 per cent in the same years. (The lowest year for foreign aid's share in expenditures was 1976; an increase in that level was budgeted for 1977.) The implication of this general trend appears to be that the Qatari government has tended to emphasise domestic expenditure in recent years and can be seen as a measure of the seriousness which the government attaches to domestic development.

Table 3.3: Qatar: Government Outlays, 1970-7[a] (QR million)

Expenditure item	1970	1971	1972	1973	1974	1975	1976	Budget 1977
Current (% of total)	358 (70.9)	475 (68.8)	672 (70.1)	746 (48.6)	1,000 (28.9)	1,737 (28.0)	1,646 (20.5)	6,888[c] (62.4)
Capital (% of total)	132 (26.1)	169 (24.5)	228 (23.8)	236 (15.4)	942 (27.2)	1,803 (29.1)	3,942 (49.1)	
Foreign aid (% of total)	—	—	27 (2.8)	370 (24.1)	522 (15.1)	892 (14.4)	306 (3.8)	431 (3.9)
Other[b] (% of total)	15 (3.0)	46 (6.7)	32 (3.3)	183 (11.9)	999 (28.8)	1,767 (28.5)	2,134 (26.6)	3,721 (33.7)
Total	505	690	959	1,535	3,463	6,199	8,023	11,040

a   See note a, Table 3.1.
b   Includes equity participation of the government in oil and other industries.
c   This figure is the sum of current and capital outlays.

Sources: IMF *Survey*, 19 August 1974, p. 258, and 15 August 1977, p. 259. The figures for current and capital expenditures were calculated upon the assumptions contained in note 11.

An alternative means to analysing the composition of government expenditures in Qatar is that contained in Table 3.3. This structure will obviously tend to generate a more fruitful analysis since there is a breakdown into current and capital expenditures; that breakdown was based on certain assumptions.[11]

Prior to 1973, the expenditure policy of the government of Qatar could be described as anti-development, since the current outlays constituted approximately 70 per cent of total outlays during that period. The situation however changed markedly in 1973 when the share of current expenditure fell below the 50 per cent mark. Nonetheless, as the table also shows, in that year the proportion of total outlays expended on capital development fell to 15.4 per cent from the average over the preceding three years of about 25 per cent. Thus capital expenditure did not benefit directly from the cutback in 1973 current spending. The reason is that the share of foreign aid and other expenditures, especially equity participation of the government in business enterprises and net lending, rose to 36 per cent from a 1972 figure of only 6.1 per cent. In the years which followed 1973, there was a significant decline in the share of current expenditure in total government outlays. Consequently, the more recent expenditure pattern of the Qatari government may be described as development-oriented; the 49.1 per cent share devoted to development in 1976 is remarkable.

As noted earlier, the proportion of government revenue devoted to foreign aid has been falling over the years in Qatar. Also in terms of its share of total government outlays, foreign aid has lost its place as a significant expenditure outlay.

Although the step-wise takeover of oil production by the government, begun in 1973 and concluded in 1977, has resulted in the rapid growth of the 'other' expenditure category in Table 3.3, it should be expected that in the future this item will be of diminished importance when the amount involved in the 1977 participation agreement has been paid in full.

A word of caution before closing the discussion on the expenditure pattern based on Table 3.3. If one interprets equity participation outlay as an investment expenditure, then the analysis of the significance of the capital item in Qatari government expenditure is likely to be different from the one above. It obviously will indicate that the extent of development orientation of government finance has been underestimated.

Pattern of Development Expenditure

Since Qatar has no formal central planning institution, the Council of Ministers fills the gap in the development process by selecting major projects. So far, the selections indicate a development strategy which emphasises capital-intensive projects using one of the most abundant resources of the country — natural gas — either as a 'feedstock or as a source of energy'.[12] Initially, the industrialisation strategy had been somewhat inward-looking aimed primarily at the domestic market. The newer policy has been to undertake export-oriented projects.

The 1977 budget allocated QR 2,197 million for industrial development. However, the 1978 budget cut down capital expenditure allocation substantially. The reduction of the capital budget was mostly accounted for by slashing the allocation for industrial development to QR 1,174 million.[13] The 1978 budget again demonstrates the ability of the Qatari government to control its spending whenever it finds that it is in the interest of the economy to do so. Table 3.4 compares the allocations for various development (or capital) items under the 1977 and 1978 budgets. To the extent that it has not been possible to separate the allocations for some of the sectors, it may not be fruitful to carry out detailed analysis based upon this table. Housing clearly received an increased emphasis in the 1978 budget. This might have been in response to the housing shortage in the country, and the resultant high rents in the city of Doha. Then, of course, the miscellaneous category labelled 'others' includes allocations for water, sewage and town planning. On the other hand, electricity, education, health, and transport and communication were de-emphasised.

Barring much quantitative analysis of the composition of the development expenditure in government finance over a period beyond two years, the only recourse is to enter into a general discussion based on secondary sources.[14] The major

Table 3.4: Qatar: Development (Capital) Budgets, 1977 and 1978

Item	%		QR million	
	1977	1978	1977	1978
1. Industrial development	34.9	22.7	2,197	1,174
2. Electricity	14.4	15.9	908	823
3. Housing	10.2	17.8	643	920
4. Transport & communication	8.9	5.3	563	271
5. Education	14.2	6.4	895	331
6. Health	6.2	2.6	391	133
7. Water & sewage		7.6		394
8. Town planning		8.8		455
9. Others	11.2	12.9	704	664
(7, 8 & 9)		(29.3)		(1,513)
Total development expenditure	100.0	100.0	6,301	5,165

Sources: Computed upon information contained in MEED, 6 January 1978, p. 27; and MEES, 24 January 1977, p. 12.

emphasis in development in Qatar to date has been to establish an infrastructural base of the economy. In addition, generous welfare programmes have been undertaken, specifically in education, health, and housing, aimed at improving the living standards of the average Qatari and as a means of redistribution of income and wealth. It is estimated that in the 1972 and 1973 fiscal years, expenditure on these and other social/welfare government activities on the average accounted for about one-third of total government outlays. Education is free for Qataris both within and outside the country. Also, non-Qataris within the country enjoy free medical services.[15]

To summarise, although there has been no comprehensive development plan in Qatar, the government so far has been able to adopt a prudent spending policy. Industrialisation is being pursued at a feasible rate; housing, education, health and other social/welfare programmes are being provided liberally; and the physical infrastructure of the country has not been neglected, although the waiting time at Doha harbour is still longer than desired. Even in the area of agriculture, the sector most ill fated because of the arid condition of the country, government spending has had beneficial effects in the form of increased productivity.

Conclusion

In conclusion, one observation should be reiterated: the fact that fiscal and monetary policies are one and the same thing. If this is so, then government policy has been not to overheat the economy. The moderate policy appears to have paid off, for inflation in Qatar has been moderate compared with what prevails in countries such as Saudi Arabia. However, as the development process continues to generate the need for more government injection of money in order to maintain the tempo of development, it will be a difficult policy decision for the government whether to reduce the rate of spending at the cost of slower development thereby gaining stable prices, or sacrifice price stability for rapid economic development. To date the Qatari government has shown the tendency to try to achieve both, perhaps because the two are not as yet strict either-or options in the economy. It may yet become necessary to have an institutionalised planning body to undertake a more comprehensive form of project formulation and allocation in the country.

Notes

1. *Financial Times*, 9 March 1977.

2. The land area is given as 4,000 square miles. Population estimates however differ from one source to another. In 1976, it was officially put at 180,000 people, including immigrants, 80 per cent of whom live in the Doha area. See *Voice of the Arab World* (London, September 1976), p. 16.

3. *Financial Times*, November 1976 to August 1977.

4. Farid Abolfathi *et al.*, *The OPEC Market to 1985* (D.C. Heath and Co., Lexington, Mass., 1977), p. 227.

5. United Nations Industrial Development Organization (UNIDO), *An Industrial Survey of Qatar* (UNIDO/TCD 103, April 1972), p. 64.

6. See Table 3.1. First oil payments to the government were made in the form of rents related to exploration activities, totalling some £20,000 (163,250 rupees) in 1939; with commercial production in 1951, revenues approached £2 million.

7. It must be pointed out that in recent years expenditure has grown very fast indeed. It may even be added that the growth has been faster than could be supported by the physical infrastructure and annual revenues.

8. The behaviour of the item 'net lending and equity participation' will be discussed later in this chapter.

9. N.A. Shilling, *Doing Business in Saudi Arabia and the Arab Gulf States* (IPIC, New York, 1975), p. 264.

10. Ibid., p. 275.

11. The assumptions were as follows: (a) that the QR 1,000.7 million 1974 budget allocation for current outlay (see IMF *Survey*, 19 August 1974, p. 258) was actually spent, and (b) the budget expenditure 1975-6 given in *Middle East Annual Review* (1977), p. 283, represented capital expenditure.

12. IMF *Survey*, 15 August 1977, p. 258. For more discussion on industrialisation and industrial strategy see the chapter on industrial and agricultural development.

13. See Table 3.4.

14. IMF *Survey*, 19 August 1974 and 15 August 1977.

15. See the chapter on social development for more discussion of the provisions of social/welfare services by the government.

4 INDUSTRIALISATION, ABSORPTIVE CAPACITY AND AGRICULTURAL DEVELOPMENT

Introduction

Even in developing countries where financial capital is scarce, the issue of whether to emphasise agriculture or industry has given way to a recognition of the interdependence of these two vital forms of economic activity. Additionally, in capital-surplus countries, such as Qatar, agricultural and industrial development are seen as areas in which the policy of diversification can be given practical manifestation. However, the interdependence of the agricultural and industrial sectors and the similar role they both play in Qatar's diversification efforts should not cloud the view of any analyst of the need to strike an optimal balance between developing agriculture and industry. Indeed, it is basically because of the interdependence and shared goals which they serve that a determined thrust should be made to ensure that the efforts in one sector are supplementary and/or complementary to development in the other.

The problem of ascertaining 'the most effective methods of allocating resources',[1] which is what the issue of balanced versus unbalanced growth is about, prevails in capital-surplus countries.[2] In these nations, the issue hinges on policies and programmes to diversify the economies. Although financial capital may not be a constraint, especially in the short term, countries like Qatar have had to face other constraints to development. Physical bottlenecks, manpower shortages, and inflation have proved to be real impediments to the absorptive capacity of their economies. In particular, the impact of these negative factors has often been most severe in the very developmental activities intrinsic to a successful diversification policy, i.e., agriculture and industry. This chapter explores the extent to which Qatar has attempted to diversify its economy in the face of these constraints. The analysis involves an examination of the rationale in the country's industrial strategy, the problem of absorptive capacity and its relationship to Qatar's industrialisation programmes, and

72

finally, the role of the agricultural sector in the nation's development.

Industrialisation in Qatar

In discussing the industrial development in Qatar several approaches can be utilised. First, the rationale for industrial development strategy now being followed in Qatar can be analysed. This is then linked with the issue of absorptive capacity. To the extent that most developing countries and, indeed, all the oil-producing states in the Gulf face the problem of absorptive capacity in varying degrees, the analysis of this problem in Qatar is undertaken within a broad, sometimes general, context. Having provided the framework within which to appraise the country's industrial development efforts, the last approach encompasses the developments which occurred in Qatar within the industrial field.

Rationale for Qatar's Industrial Strategy

As noted earlier, with an economy which depends on oil for 90 per cent of its foreign exchange earnings,[3] Qatar has certain special obstacles and opportunities for development. According to the Heckscher-Ohlin theory of international trade, a country tends to have lower comparative costs in the commodity that uses the largest amount of the relatively cheapest factor in that state.[4] This theory serves as the basis for specialisation. In this thinking, Qatar should establish and promote industries and types of activities which are best tailored to suit the quantity and quality of the country's natural and human resources.

Qatar has ample reserves of high-gravity, low-sulphur crude oil, which are the source of substantial capital funds. There are, as well, tremendous quantities of natural gas, until recently completely flared. It follows that industries primarily based on natural gas and/or oil are most likely to be efficient and successful. Fortunately, these industries are not only energy intensive but also capital intensive. Thus, they tend to reduce the total manpower requirement in the economy and assist in the process of Qatarisation of labour and population. Within the Qatari oil industry itself, fewer than 2,000 persons are employed, including the offshore fields.

The petrochemical industry is an example of an under-taking which is both capital and energy intensive.[5] The products of this industry consist of basic or first generation petrochemicals, and intermediate and end products or second generation petrochemicals.[6] Plastics, which belong to the latter category, must be converted into various end products before classification as consumer or industrial level goods. The petrochemical industry enjoys certain principal charac-teristics of which some facilitate while others hinder its establishment in Qatar. The specific features making the petrochemical industry attractive include the nature of its inputs of natural gas and oil and the high capital require-ments.

The petrochemical industry requires a very high capital/labour ratio. In fact, investment per new job created is estimated at $20,000 to $100,000.[7] Moreover, as the basic products and intermediates advance from the stage of trans-formation to finished products and then on to consumer goods, larger investments are needed. The investment required for the transformation of finished products into consumer or industrial goods (the third manufacturing phase) is two to three times higher than that required for the production of intermediate products and the latter's transformation into finished products (second manufacturing phase), and five times higher than that necessary for the production of basic products (first manufacturing phase).[8] Therefore the establishment of petrochemical plants in Qatar is one way to enable the country to industrialise without exerting un-warranted pressure on its most scarce input, labour. In addition, the petrochemical industry would tend to expand gas utilisation, thereby turning a largely wasted resource into a base for economic activity.

Those features of the petrochemical industry which may pose some difficulties in the immediate future are related to two factors. First, the industry is characterised by sophisti-cated technology, along with a high level of competition. Installations must utilise the most modern technology and highly skilled manpower. Moreover, the industry is exceed-ingly dynamic and research into new processes and products plays a crucial role. Competition remains intense due to the dynamic nature of the industry traceable to the availability of various substitutes. Further, the different items produced by the industry compete with each other, as cellophane, waxed paper, aluminium foil, polyethylene sheets, and acetate film — all have comparable uses. Currently, Qatar lacks adequate trained, indigenous manpower necessary for the petrochemical industry; thus importation of labour is necessary.

The second factor which militates against the establishment of such an industry is the fact that production caters almost exclusively to the export market. What complicates the marketing problem is that several Gulf states and other Middle Eastern countries, with similar petroleum resources as Qatar, have constructed or are contemplating the establishment of petrochemical complexes. To avoid problems of underutilised plant capacity or oversupply of petrochemical products, both of which tend to reduce the profitability of these ventures, some coordination and cooperation between these nations seems to be an indispensable condition for success. In fact, the probable benefits of some level of economic integration between these countries transcend the petrochemical sector, perhaps to involve larger portions of the economy of each state.[9]

Other capital-energy-intensive industries are aluminium and iron smelting. These enterprises, together with the petrochemical groups, could constitute the backbone of the future industrial sector in Qatar. The availability of raw materials necessary for the production of cement also makes such industry feasible. An additional diversification venture that has future prospects and is completely unrelated to oil and natural gas is the development of the fishing industry. The objective would be not only to produce fresh fish, but to establish and expand fish canning and other final products which have fish as the primary input. Prospects for a very significant expansion in agricultural production are not particularly favourable due to the scarcity of fertile land and the inadequacy of water resources. Consequently, agriculture-based manufacturing plants generally may be undertaken only on a small scale.

As a capital surplus country, profitable regional investment ventures are to Qatar's advantage. In fact, in coming years the Arab states of the Gulf, including Qatar, can develop together into a financial centre that may serve not only the Gulf region, but the entire Middle East, North Africa, and perhaps other areas of the world.[10]

While Qatar has been spared the problem of inadequate financial capital which confronts most developing countries, two factors, however, represent major barriers to economic development in Qatar: the small size of the domestic market and the limited absorptive capacity of the economy. These problems are not unique to Qatar but are shared by some other Gulf states as well.

The narrowness of the market (which is also an integral part of the absorptive capacity issue) primarily results from the smallness of Qatar's territory and a population of around 200,000. Such a population base cannot generate sufficient local demand for most modern, large-scale industries. Industries catering to the domestic market therefore must necessarily be small-scale and/or cottage industries. Moreover, the market is further restricted by the fact that the majority of the population consists of non-Qataris who often remit a considerable portion of their earned incomes to their home countries. This reduces the effective demand for local goods and services, and diminishes the multiplier impact of increased government expenditures and private investment outlays.

Although small-scale and/or cottage industries are recommended, these industries can be hindered by a liberal trade policy as in general they are not able to compete with the modern, large-scale foreign enterprises.[11] Nonetheless, the establishment of industries behind high tariff walls may not be advisable when such a policy reduces the efficiency of the economy and adversely affects income distribution.

The Absorptive Capacity Problem

In the drive for diversification and for a greater degree of balance in the economy, Qatar is confronted with an acute and serious problem of factor imbalance. On the one hand, capital funds from oil revenues are abundant and rising; conversely the co-operant factors of production of land, labour, and technical skills, are both qualitatively deficient and not expanding at a rate commensurate with the growth in oil revenues. The availability of capital can correct or help substitute for certain deficiencies and scarcities in the co-operant factors of production. Nevertheless, the process of substituting capital for these factors is not boundless and is subject to the law of diminishing returns. The extent to which capital can be used productively depends primarily on the absorptive capacity of the economy.

Absorptive capacity is defined by development economist Benjamin Higgins as 'the amount of investment that can be undertaken, over a five-year planning period, beginning from the present, without reducing the addition to perpetual national income below three percent'.[12] Evidently the 3 per cent is arbitrary and reflects basically Higgins's estimation of the opportunity costs of the committed funds.

Branko Horvat,[13] however, believes that the absorptive capacity of an economy is only reached when the marginal efficiency of social investment is zero. Horvat's definition is rather subtle. It does not imply that the absorptive capacity would be reached when an additional private investment is earning zero returns. On the contrary, additional investment is likely to remain profitable from the point of view of the individual; but from society's viewpoint it might contribute nothing to the national product, since it generates so many external diseconomies that its beneficial effects are neutralised completely. Horvat, moreover, emphasises the role of investment in expanding the absorptive capacity of the economy.

John Adler, an economist well known for his work in this area, stresses that:

Absorptive capacity may then be defined as that amount of investment, or that rate of gross domestic investment expressed as a proportion of gross national product, that can be made at an acceptable rate of return with the supply of co-operant factors considered as given. This is not to say that the investor, or the investing authority, would not attempt to increase the supply of co-operant factors. But, in the short-run, this increase is either a physical impossibility or is so costly that it increases sharply the cost of investment or the total operating cost, and thereby reduces the return on capital below the acceptable rate.[14]

So far these definitions emphasise the socially acceptable cut-off rate or the socially admissible rate of return on invested capital. Kindleberger, however, takes a different approach based on the availability of certain cooperating factors of production without paying explicit attention to the current rate of discount. Thus, 'with limited economic engineering, and administrative talent, a developing country can prepare so many projects and no more'.[15]

The capacity of the economy to absorb capital depends primarily on the human factor. This element is very complex and perhaps may be made a function of four basic factors (policy variables) and of their rates of change. The four are: personnel consumption (C); health (H); knowledge (Kn); and economic and political organisation (O). In addition, the remaining relevant features of the economy could be lumped together conveniently as a single factor (E).[16]

The limitations on absorptive capacity are varied and may take many forms, 'and it is not very meaningful in practice to propose policies to increase absorptive capacity in general. The only way to come to grips with the practical limitations of absorptive capacity is to devise specific measures to raise specific limitations'.[17]

Some of these specific limitations[18] pertain to the lack of knowledge concerning natural resources or the availability of technology. Another serious constraint is the lack of skills necessary (a) to prepare investment projects, that is, to study the engineering, economic, and financial aspects of a project, (b) to implement investment projects once their feasibility is determined, and (c) to perform the manufacturing and other related tasks of the new enterprises. Adler also emphasises a third limitation, namely, the lack of managerial talent and experience, as a separate factor distinct from the general lack of skills, noting: 'the confusion between technical skills and managerial competence, frequently observed in less developed countries, is one of the prime causes of the low rate of return from state economic enterprises and thus limits absorptive capacity'.[19]

Another set of curbs, but at the level of the enterprise itself, consists of institutional, cultural and social constraints. Institutional limitations can include the maintenance of stability, law, and order, the nature of administrative procedures, the degree of communication and coordination between government departments, and the overall efficiency of the governmental machine. Cultural and social constraints refer to differences in the social structure and the cultural values between advanced and developing countries. Institutional limitations along with the cultural and social constraints, unlike the other restraining factors mentioned above, 'are not directly amenable to technical assistance or concerted action. They can be overcome only by the process of development itself'.[20]

In the light of the previous discussion, one may ask what Qatar's limited absorptive capacity can be attributed to. Is it because a rigid definition of absorptive capacity applies, i.e., one that requires a high social rate of return? Or is it perhaps due to the low level of the factors determining absorptive capacity which are also not expanding at a sufficiently rapid rate?

To the extent that a project is evaluated in isolation or semi-isolation, the expected rate of return would not reflect the social rate of return of the respective project.[21] In general, this method of evaluation would underestimate the project's expected rate of return and thus limit absorptive capacity. Additionally, the United Nations Industrial Development Organization (UNIDO) argues that unless a prospective project has a rate of return of at least 7 per cent, it should not be established, since this rate of return is generally obtainable from investment abroad.[22] It could be noted that a 7 per cent rate of return might be considered rather high because the process of industrialisation in Qatar is in its infancy, and that investment abroad should not be guided only by the yield it produces but also by the elements of security and liquidity of the committed funds. Moreover, it must be remembered that investment abroad does not generate the non-economic benefits usually associated with domestic investment. These spin-off factors, such as national pride, self-reliance, and the fulfilment people derive from productive work, are extremely important to the government which is concerned with the welfare of its citizens. Thus, the government might adopt Horvat's definition of absorptive capacity as it permits the maximum amount of productive domestic investment to be undertaken.

With regard to Qatar's population, the composition and characteristics of the labour force indicate that the human factor, which is the major determinant of absorptive capacity, constitutes a real bottleneck constricting the extent and pace of development. The availability of capital can, and thus far has, abridged the difficulties associated with a limited supply of labour in general and of skilled and highly skilled labour in particular. The continued importation of labour, though it increases the absorptive capacity of the economy, has the adverse effect of increasing the proportion of non-Qataris in the total population. Thus, the government is faced with a trade-off between rapid industrialisation and diversification of the economy and the goal of Qatarisation. Here, as elsewhere, a balance could be struck between two competing ends.

Industrial Development in Qatar

The government of Qatar, as with other petroleum-based economies in the Gulf, accords industrial development top priority among its efforts to diversify the economy and to reduce the almost complete dependence on the export of crude oil as a source of national income. In 1977, the planned expenditure on industry was 34.9 per cent of total government project allocation; although the governmental policy to reduce the tempo of economic activity in Qatar and thereby arrest inflationary tendencies caused the 1978 allocation for industry to be cut, it still constitutes 22.7 per cent of total projected investment.

Several projects were initiated in the period prior to 1975 as part of the Umm Said industrial complex. Table 4.1 gives estimated allocations for various aspects of this complex, and for the Umm Bab area, where the cement factory is located. Since then, work on three more industrial projects has either been started or their agreements or protocols signed. These are (a) an iron and steel plant, (b) a petrochemical plant in Qatar, and (c) a petrochemical plant located outside Qatar, in Dunkirk, France. In addition, fertiliser and other plants are to be expanded. Further evidence of the extent of the government's involvement with industrial development is shown in Table 4.2 which presents the allocations for several projects.

Table 4.1: Estimates for Major Capital Investments in Industrial Projects, 1971-5[a]

Project	QR million
Flour mill	3.8
Qatar Fertilizer Co. (QAFCO)	274.1
QAFCO extension	251.3
Refinery	38.0
Liquid petroleum gas[b]	228.4
Solar salt	15.1
Bag plant	3.0
Plastic forming[c]	0.5
Total for Umm Said area[d]	814.2
Cement at Umm Bab	37.7
Cement extension	46.6
Total investment	898.5

[a] Excludes any capital charges.
[b] Includes jetty and storage tanks.
[c] Includes blow-moulding, injection and extruders.
[d] Expenditure by the government for the provision of infrastructure in this area is in the order of QR 60-80 million.

Sources: UNIDO, *Industrial Survey*, p. 32; *Qatar into the Seventies* (Ministry of Information, Qatar, 1973), pp. 96-8.

In 1977, fire swept through the Umm Said natural gas liquefication plant. Before that catastrophe, industrial development and diversification were moving forward in Qatar at a rapid pace. Official circles felt the country was years ahead of any other Gulf state in this area of economic endeavour. Indeed, processing facilities for petrochemical products were in operation before the fire, and at the time operations began, few people would doubt the optimism of the Qatari authorities that by the end of the present decade 'Qatar would be the first state in the Gulf to stop flaring gas altogether'.[23] At present, however, the damage done by the fire has placed considerable pressure on those responsible for realising this objective. An alternative course which is likely to be taken for the time being is to sacrifice rapid progress in order to hold down the inflation which physical and man-power bottlenecks are likely to generate together with increased spending.

One of the early oil-based industrial projects established in the country is the fertiliser plant managed by the Qatar Fertilizer Company (QAFCO). This firm, which was set up in 1969 and is located in the Umm Said industrial area, is a joint venture with capital of QR 57 million subscribed as follows: Qatar, 63 per cent; Norsk Hydro of Norway, 25 per cent; Hambros Bank of the United Kingdom, 5 per cent; and British Power Gas Corporation, 7 per cent.[24] Although the general structure of ownership of 63 per cent Qatari interest and 37 per cent non-Qatari interest has remained the same since its inception, the share of Hambros has dropped from its original allotment of 10 per cent while Norsk Hydro has gained by 15 percentage points. The foreign partners have very crucial inputs in the construction, management, and operation of the fertiliser facility, and in the marketing of its products. There is a good working partnership and mutual cooperation between the government of Qatar and these international firms.

The raw material and primary energy source for the fertiliser complex is provided by a pipeline which taps the natural gas at the Dukhan oilfield, 80 miles away. However, all of the secondary electric power and water needs of the plant are provided by its desalination plants. The technical operation of the fertiliser plant has been described thus:

> The factory operates on natural gas and uses light natural gases with a high methane content as raw material. After being processed gases are taken to the ammonia units which have a production capacity of 900 tons of ammonia, 600 tons of which is used to produce 1,000 tons of urea fertilizer per day. The remaining 300 tons of ammonia are kept in a special cold store ready for export.[25]

Some of the urea is exported while part is utilised in Qatar to promote agricultural production. The growth of the fertiliser industry is believed to be the cornerstone of the country's agricultural boom in recent years.

An expansion of the capacity of the fertiliser factory is

Table 4.2: Major Industrial Allocations in Qatar, 1975-7[a] (QR million)

Item	Estimated total cóst	Allocation for 1975	Allocation for 1976	Allocation for 1977
Iron and steel complex	1,200	50	275	325
Petrochemical complex	2,100	100	250	365
Natural gas complex	1,400	60	350	290
Cement factory extension	800	75	200	n.a.
Fertiliser plant extension	1,000	n.a.	350	250
Umm Said town development	1,500	n.a.	164	n.a.
Umm Said electricity supply	1,700	n.a.	350	n.a.

[a] Estimates given in dollars have been converted into Qatar riyals at the exchange rate 1 QR = $0.25.

Sources: *Bulletin of the American-Arab Association*, March 1976, p. 3; MEED, 21 January 1977, p. 28.

under way. When this is completed, the output could reach 18,000 tons of ammonia and 2,000 tons of urea per day. The cost of the extension is estimated to be about QR 1,000 million. Work on the project began in 1975. As Table 4.2 indicates, the projected government expenditures on the fertiliser plant extension in 1976 and 1977 are QR 350 million and QR 250 million, respectively.

QAFCO employs about 540 persons. They include 40 top, middle and lower middle personnel, 250 supervisors, plant operators, clerks, and craftsmen, and 250 unskilled workers. The completion of the extension on the plant would also necessitate additional workers. In fact the original design included a provision for expansion to be implemented in response to the expanded market for the factory's products. It was then estimated that should such an expansion be made, an additional 175 workers, mostly in the skilled and semi-skilled categories, would be required.[26]

The expansion of the plant has been found feasible despite the earlier fear that increased fertiliser production in Iran, Kuwait, Iraq, and other Middle Eastern countries would make such expansion impossible unless contractual agreements between major consuming countries and Qatar preceded such a step. The extension to the existing fertiliser plant has been given the go-ahead signal, although it is believed there is little likelihood the plant will be run at more than 80 per cent capacity for some years. The management of the plant and the government are optimistic because of the current boom in the international fertiliser market.

Perhaps a more important institutional arrangement would involve cooperation among Middle Eastern fertiliser-producing countries in order to avoid a glut in the international market for that commodity. This suggested line of action could be investigated for implementation vis-à-vis almost all products which will come out of the industrialisation programmes based on oil; the availability of the primary input into petrochemicals makes these countries produce identical products. Thus, as noted earlier, a cooperative effort on the part of all states concerned is inevitable for maximising the future prospects of the petrochemical industry within this region.

The government's contribution to the industrial development in the Umm Said area transcends sharing the necessary capital for the various joint ventures. In addition to the equity participation, the Qatari government has spent a large amount of money on the provision of infrastructural facilities for industry. For the fertiliser complex in particular, facilities such as roads, loading harbour and pipelines, among others, are available and are estimated to have cost the government about 30 million pounds sterling (or approximately QR 204 million at the 1976 rate of exchange of £1.00 = QR 6.8). QAFCO would benefit as well from the Umm Said town and power development projects, estimated to cost the government QR 1,500 million and QR 1,700 million, respectively.[27]

The massive fire at the Umm Said natural gas liquefication plant on 3 April 1977 has obviously been a setback in what appeared to be systematic progress toward effective utilisation of Qatar's natural gas resources. The liquefication unit was part of a complex scheme to harness the gas from the offshore oilfield of Dukhan.[28] The plant, which was opened in June 1975, consisted of a gas collection and an initial processing plant located close to the Dukhan oilfield. The same pipeline which serves the fertiliser complex was the means for transporting gas to the factionation processing plant at Umm Said. The Umm Said unit has the capacity to produce 350 million cubic feet of feed gas a day, and in addition, 20,000, 8,000 and 4,000 barrels a day of propane, butane, and natural gasoline, respectively. The plant was taken over by the Qatar Petroleum Producing Authority (QPPA) in 1976 after its construction by Shell for the Qatar Petroleum Company.

Another development in the Qatari petrochemical industry was the formation of Qatar Gas Company Ltd in December 1974. This is a joint Qatar-Shell venture. The Qatar government owns 70 per cent of the QR 385.6 million capital while a Shell subsidiary, Shell Gas BV, contributed the remaining 30 per cent of the capital. The proposed activities of the company include investing in gas projects, trading, and transportation of natural gas. The NGL plant of the Qatar Gas Company is scheduled to go into operation by 1978; however, the recent fire may postpone the schedule.[29]

Despite the setback which the fire disaster might have brought to the petrochemical industry, efforts at continued industrialisation in this sphere are largely undaunted. The 1977 budget allocated QR 290 million to the construction of a new NGL plant. The Mitsubishi Corporation and Chiyoda Chemical Engineering and Construction Company, both from Japan, have the responsibility of building the new plant known as NGL-2. The new plant will be fully Qatari-owned (by Qatar General Petroleum Company — QGPC). The principal market for the new plant will be in Japan.[30]

The most expensive of all the projects envisaged in Qatar is the petrochemical plant being built by the Qatar Petrochemical Company (QAPCO). On 27 June 1974, an agreement for the establishment of QAPCO was signed between the government of Qatar and two French firms, CdF Chimie and Gazocean Company. The former is an experienced petrochemical company while the latter specialises in the transportation of liquid gases. The agreement was consummated in Royal Decree No. 109 (1974), establishing the petrochemical company with a capital of QR 2,400 million with subscription shares of: Qatar government, 80 per cent; CdF Chimie, 15 per cent; and Gazocean, 5 per cent.[31] In addition, two subsidiary companies were to be established. For the establishment of the first, an agreement was signed between the Qatar government and Gazocean. It provided for the setting up of a joint company owned 60 per cent by Qatar and 40 per cent by Gazocean, with a capital of QR 50 million. The company was to be responsible for transporting petrochemical products and gas liquids. Further, it was envisaged that a marketing and distribution company jointly owned by Gazocean and Qatar would be established to market ethylene. A second agreement was also signed with CdF Chimie 'for the supervision and management of and for marketing of polythene'.[32] However, Gazocean left the original petrochemical company agreement; presently, the Qatari government has 84 per cent and CdF Chimie 16 per cent of the capital.[33] The 1976 and 1977 budgets have allocated 250 and 365 million riyals, respectively, to the project.

Interestingly, the proposed French/Qatari petrochemical venture has another sector to be located at Dunkirk in France, which is to be owned 40 per cent by the Qatari government. The Dunkirk project is under the control of the North Company for Petrochemicals Production (NCPP) which has a capital of 100 million French francs (or approximately QR 70 million).

Perhaps at this stage it would be helpful to clarify the organisational structure of the petrochemical industry in Qatar from the point of view of control. The general supervision of QAFCO, the country's two oil and gas operating companies, the distribution company, Qatar Gas Company, the Qatar sector of the French-Qatari ventures, Qatar's share of the Bahrain dry dock belonging to OAPEC (the Organization of Arab Petroleum Exporting Countries), and other OAPEC concerns to which Qatar has contributed, including pipelines, all are in the hands of Qatar General Petroleum Corporation (QGPC), set up in 1974.[34] The QGPC was established as a profit-seeking enterprise to be run along commercial lines, with a capital of 1 billion riyals (which was later doubled). In practice, the direct control over its domestic production operations is the prerogative of a subsidiary, the Qatar Petroleum Producing Authority, formed in 1976.

There are two major non-oil-based industrial projects in which the Qatari government participates. These are the iron and steel projects begun in 1975 and the cement factory which has been in production since 1969. Whereas the cement factory is owned jointly by the Qatari government and the local private sector, the Qatar Iron and Steel Company Ltd (QISC) is a joint venture between the government and two Japanese companies — Kobe Steel Limited and Tokyo Boaki.

Although the capital of the QISC is QR 200 million (Qatar having 70 per cent participation), the total cost of the iron and steel complex is projected to be in the region of QR 1,600 million. Thus, like the petrochemical industry, external loans are going to be relied upon to finance most of the project. The Kobe Steel Limited, which owns 20 per cent of the complex, has the production contract, while Boaki (with 10 per cent of the ownership) is responsible for marketing the product. Like most joint ventures with foreign companies, it appears that the Qatari government has been able to secure the services of companies which have wide experience in their respective fields and has entered into contracts with them.

While almost all of the iron ore and scrap needed to feed the plant will have to be imported, the availability of natural gas for use in the reduction stage and to generate electrical power to feed the complex makes the venture an attractive one. However, the amounts of iron ore and scrap to be imported have created some scepticism in certain circles. The production capacity of the plant initially will be about 400,000 tons of ferroconcrete per year, less than 20 per cent of which will be sufficient to meet domestic demand, with the balance available for export. Despite the pessimism just noted, officials believe that production will start in 1978, and that the project will be able to recoup the amount expended on it within six years. The importance which the government of Qatar attaches to this project is shown by the allocations for it in the 1976 and 1977 budgets of QR 275 million and QR 325 million, respectively.

The Qatar National Cement Company was formed in 1965 to utilise the limestone and clay available in great quantities at Umm Bab. The establishment of the factory constituted the first major industrial venture in Qatar; it was also meant to meet rapidly increasing demand for cement in the construction sector. However, since Qatar does not have gypsum or iron ore, these raw materials are imported from Saudi Arabia. Work on the factory commenced in early 1967; actual production began in May 1969. By the end of that year, the output of portland cement by the 'dry process' method had exceeded the designed initial plant capacity of 109,500 tons per year.

In 1974 and 1975 an extension was begun on the factory which included the construction of two more furnaces, thereby raising its capacity to 326,000 tons per year. The extension was in response to the increased need for cement which has also led to the decision to build a fourth furnace to raise the capacity of the factory to 600,000 tons annually, in order to satisfy future needs.[35] Indeed a recent official estimate puts the ultimate production target of the cement factory at 566,000 tons per annum.[36] The entire cement complex (including extensions) is expected to cost in the neighbourhood of QR 800 million: the Qatari authorities expect it will pay for itself.

This discussion thus far has been of industrial projects in which the government participates. There are of course a host of small private businesses for which data are not available. However, a particular private industrial concern worth mentioning is the flour mill run by the Qatar Flour Mills Company. The plant, located adjacent to QAFCO in the Umm Said industrial area, has silos with a capacity of 8,000 tons to serve as a buffer stock to isolate the country from excessive volatility in world wheat prices and to ensure a regular supply of wheat for grinding at a convenient and economical rate throughout the year. The flour mill, completed in September 1972, started with the initial output of 100 tons per day.

Other areas in which the private sector participates, with the encouragement of the government (which in some cases provides the infrastructural framework), are plastics, detergents, paints, and paper. Even in light industries, the Qatari government recently has shown increased direct interest. The Industrial Development Technical Center (IDTC) which is not only the principal adviser to the government on economic policy but handles all of Qatar's new industrial projects, including feasibility studies, is reported to favour more light industries for the country. It is felt that the government has now initiated all the feasible major heavy industrial projects; it is time to turn attention to light industries.[37] The IDTC has commissioned a French firm, Serete of Paris, to carry out 13 feasibility studies of light industries including food, chemical derivatives, building materials, and plastics. From these studies a light industry development plan will be drawn up, which, when approved, will form a second blueprint for the industrialisation of Qatar.[38] The indication that the Qatari government may be shifting towards light rather than heavy industry is perhaps more clearly demonstrated by the decision to abandon the projected aluminium smelter for the moment; this might also be due to the manner in which negotiations on the project were proceeding, falling below the government's expectations.

Agricultural Development

A nine-point agricultural development programme was initiated in Qatar in 1970 under the responsibility of the Ministry of Industry and Agriculture. The main policy objective is to increase the output and the quantity of livestock, fruits and vegetables so that the country becomes largely self-sufficient in food and other farm products. Other areas considered under the programme are research, land reclamation, regulation of pasture, the establishment of training centres, hiring of qualified personnel, purchase of necessary production materials, and cooperation with other countries and international organisations in the fight against agricultural pests. Since the policy was instituted, production has increased rapidly in spite of the natural limitations of aridity and soil. Arable land and irrigation water are scarce. The soil consists mainly of sandy loams, rarely exceeding two feet in depth. However, scattered tracks of land suitable for agriculture are available and are primarily concentrated in the northern half of the country. Also, there are projects for forestation, experimental farms such as Rodait al Khail, vast irrigation schemes, and assistance to farmers.

The United Nations Food and Agriculture Organization (FAO) has assisted the Qatari government in an extensive hydrological survey to ascertain the agricultural potential of the country. To help with the nine-point programme, the FAO had a hand in the establishment of training centres and in research. As a result, a spring barley strain has proved successful, and a suitable variety of wheat is currently being developed. The first experimental farm began operation in

1963; the 1976 budget allocated QR 5 million for such farms.[39]

The climatic and soil conditions have created an obstacle to development, along with the supplementary problem of attempting to settle the nomadic bedouins in farming communities. It is here that the training centres are relied upon to deal with a socio-economic human problem.

Since World War II, emphasis in the emerging nations has been on industrial development rather than on the agricultural sector. The prestige seen as reflected from having an 'advanced' sector (that is, an industry along the lines obtaining in Western economies) and pressure from the indigenous population desiring rapid economic improvements tend to force governments in the Third World into accelerating industrialisation. The success of Qatar's drive toward self-sufficiency in food products thus depends to a large measure on the government's ability to encourage workers into this sector. The farm workers numbered only 1,360 in 1973 (of a population of 170,000), with 17,469 tons of vegetables produced. In 1974, vegetable production had increased to 18,340 tons (see Table 4.3), on 450 farms with an average holding of 30 dunums. It is even estimated that Qatar will achieve self-sufficiency in food by 1980; the country already exports fish.[40]

In 1968, the Ministry of Industry and Agriculture announced that domestic vegetable output was then meeting all of Qatar's produce consumption needs. In 1975, exports of surplus vegetables to neighbouring states totalled some QR 1 million. However, in the same year, the food import bill for Qatar was QR 200.9 million, indicating that the overall goal of self-sufficiency in foodstuffs had not yet been reached.[41]

The government, in an attempt to encourage agricultural development, provides farmers with necessary services at no charge. These include the distribution of seeds or saplings, the provision of insecticides, organising pest control, and carrying out tractor ploughing which utilises the government's own fleet of equipment. The private sector is responsible for agricultural and horticultural production, although cooperative farms are encouraged. Thus far, QR 16 million has been allocated for the development of such farms.

On the recommendation of the FAO, in 1975 the govern-

Table 4.3: Cultivated Area and Agricultural Production

Year	Vegetables (tons)	Farms	Fodder production in tons	Number of workers	Dunums[a] under cultivation
1960	1,708	119	750	606	854
1965	15,360	284	2,020	n.a.	n.a.
1973	17,469	450	n.a.	1,360	9,413
1974	18,340	450	25,520	n.a.	n.a.

[a] One dunum equals approximately 1,000 square metres or about one-quarter of an acre.

Sources: MEED, special report, April 1977; *Middle East Annual Review*, 1977.

ment established a poultry farm at Umm Qarn about 30 miles north of Doha at a cost of nearly QR 5 million. In the future, the poultry farm is expected to meet up to 80 per cent of the domestic demand.[42] Another recent undertaking is the development of a dairy farm in the private sector. The Qatar Dairy Company has imported Friesian cows from Australia to produce more than 3,000 litres of milk a day. In addition, a sheep-rearing project has been established in southern Qatar to support as many as 13,000 head of sheep. Lamb, of course, is the most widely consumed meat in the Middle East.

Total food imports in 1975 amounted to QR 200.9 million ($50.8 million) which accounted for 12.5 per cent of the value of all imports. The major items were sheep and goats from Somalia and Australia worth QR 43 million ($10.8 million). Another large item was frozen chicken from China. Qatar thus relies heavily on imports, and with the continuous rise in population and the high level of per capita income, large quantities of meat and meat products are in high demand. The Qatari authorities have made great strides in their ambitious development programme through emphasis on the agricultural sector and the relatively large sums of money allocated to it.

The fishing industry, which has always been an important area of production for the Gulf states, is similarly being considered by the government. Along with other Gulf countries, Qatar is engaged in a three-year study at a cost of $7 million to identify the major fishing resources which can be exploited. The FAO also has conducted a survey in the Indian Ocean to determine long-range possibilities. The major institution in the fishing industry is the Qatar National Fishing Company (QNFCO), incorporated in Doha in May 1966 to fish for shrimp in Qatari territorial water under a concession due to expire on 31 August 1982. Of the total capital in the company, 60 per cent is contributed by the government and the balance is owned by Ross International, a subsidiary of Imperial Foods in the United Kingdom. This enterprise is considered the most successful of the sector, having sales of QR 5 million and declaring a 20 per cent dividend in 1976.[43] Most of the output, frozen shrimp, is marketed in the United States and Japan. The company has six vessels designed for use in shallow waters, along with a refrigeration plant and processing equipment.

The government of Qatar appears to be very much committed to ensuring increased production in agriculture. Fortunately for the country, the agricultural sector has been able to attract labour despite the overall crunch in the Qatar labour market, quite the contrary to the trend obtaining in most developing countries. The government's large financial contribution should not be interpreted as 'a financial and ostentatious way of spending money',[44] but rather as a genuine effort on its part to make the country self-sufficient in major food items. This would be a vital contribution to the diversification policy in an economy where oil's eventual depletion must figure prominently.

Conclusion

To sum up, we can see that Qatar is aiming at a serious reduction in its economy's dependence on oil. Indeed, the gas-related industrial projects hold the best prospect for removing this reliance since a large deposit of reserves of non-associated gas is available for tapping. In the iron and steel industry as well, its utilisation of gas makes its prospects independent of crude oil.

In the agricultural sector, the efforts being made both by the government and by private sectors are commendable. That a country as arid as Qatar has been able to attain self-sufficiency in vegetables and is likely to repeat this performance for most food items by the end of the present decade is by no means a small achievement. It is even more remarkable when it is recognised that as recently as 1954, modern agriculture was virtually non-existent in Qatar.[45] Therefore, although for some years to come Qatar's industrial scene will continue to be dominated by oil-based industries, and the economy will be dominated by oil even further into the future, it is unlikely that, when exports of crude ultimately dwindle, either the industrial wheels of the country or the economy itself will grind to an abrupt halt.

Notes

1. Ragaei El Mallakh, *Economic Development and Regional Cooperation: Kuwait* (University of Chicago Press, Chicago, 1968), p. 101.

2. For a good collection of discussions on balanced versus unbalanced growth theories, see G.M. Meier, *Leading Issues in Economic Development*, 3rd ed. (Oxford University Press, New York, 1977).

3. IMF *Survey*, 15 August 1977, p. 258.

4. Delbert A. Snider, *Introduction to International Economics*, 4th ed. (Richard D. Irwin, Inc., Homewood, Illinois, 1967), pp. 34-42.

5. See Claude Mercier, *Petrochemical Industry and the Possibilities of Its Establishment in the Developing Countries* (Editions Techniques, Paris, 1966), for more information regarding the petrochemical industry.

6. Basic or first generation petrochemicals include the following: olefin, acetylene, butadiene, aromatic hydrocarbon, ammonia, and methanol. End products or second generation consist of plastics (polyolefins, vinyls, polystyrene), synthetic fibres (nylon-type, acrylic, polyester), synthetic rubbers (styrene-butadiene, streo-regular), detergents, and nitrogen fertilisers.

7. C. Mercier, *Petrochemical Industry*, p. 7.

8. Ibid., p. 11.

9. See the chapter on international trade and regional cooperation for a further examination of economic integration among the Gulf countries.

10. Elizabeth Monroe, ed., *The Changing Balance of Power in the Persian Gulf* (The American Universities Field Services, Inc., New York, 1972), pp. 32-4. See the chapter on international trade and regional cooperation in this volume for greater detail on this aspect.

11. Chapter 3 (public finance) outlined the issue of import duties to protect local industries.

12. Benjamin Higgins, *Economic Development: Problems, Principles and Policies*, rev. ed. (W.W. Norton and Co., Inc., New York, 1968), p. 581.

13. Branko Horvat, 'The Optimum Rate of Investment', *Economic Journal*, December 1958.

14. John H. Adler, *Absorptive Capacity: The Concept and its Determinants* (The Brookings Institution, Washington, DC, 1965), p. 5.

15. Charles D. Kindleberger, *Economic Development*, 2nd ed. (McGraw-Hill Book Company, New York, 1965), p. 326.

16. The relationship between absorptive capacity (A) and these factors may be expressed mathematically as follows: $A.dA/dt = F(C.dC/dt, H.dH/dt, Kn.dKn/dt, O.dO/dt, E.dE/dt)$. Vide: Branko Horvat, 'The Optimum Rate of Investment',

p. 751. Note that absorptive capacity is an increasing function of the factors. Also dX/dt is the time derivative (or rate of change) of factor X. X, of course, refers to either A, C, H, Kn, O, or E. Thus, for example, the term C.dC/dt represents the product of personal consumption (C) and its rate of change (dC/dt), i.e., the change in personal consumption during the time period analysed.

17. John Adler, *Absorptive Capacity*, p. 31.

18. See discussion in ibid., pp. 31-4.

19. Ibid., p. 33.

20. Ibid., p. 34.

21. Appraising a project in isolation means ascertaining the rate of return a project would yield irrespective of the external economies or diseconomies it may produce. Semi-isolation appraisal would consider the impact of the project on the extant surrounding projects. However, the possibility that profitability may be enhanced if a larger and more diversified scale of investment is undertaken is not considered.

22. United Nations Industrial Development Organization (UNIDO), *An Industrial Survey of Qatar* (UNIDO/TCD 103, April 1972), pp. 10-11.

23. N.A. Shilling, *Doing Business in Saudi Arabia and the Arab Gulf States*, 1977 Supplement (IPIC, New York, 1977), p. 77.

24. *Qatar News*, September-October, 1977, p. 7; *Qatar Fertilizer Company (S.A.Q.)* (QAFCO, Umm Said, February 1973).

25. *Qatar News*, September-October, 1977, p. 7.

26. UNIDO, *An Industrial Survey*, p. 74.

27. See Table 4.2 above.

28. MEED, 15 April 1977, p. 3.

29. MEES, January 1975, p. 6.

30. MEED, special report, April 1977, pp. 10-11. Although most of the produce will be headed for Japan, it is estimated for the future that possibly 600,000 tons annually of liquefied gas will have to marketed elsewhere.

31. *Qatar News*, September-October, 1977, p. 6.

32. Ibid.

33. *Middle East Annual Review 1978* (The Middle East Review Co. Ltd, Essex, England, 1977), p. 303.

34. Ibid., p. 307.

35. MEED Annual Report, April 1977, p.11.

36. *Voice*, 16 September 1977, p. 16.

37. MEED, 9 September 1977, p. 31.

38. MEED, 3 June 1977, p. 33.

39. MEED, special report, April 1977, p. 14.

40. N.A. Shilling, *Business in Saudi Arabia*, p. 260.

41. MEED, special report, April 1977, p. 14.

42. *Middle East Annual Review 1977* (The Middle East Review Co. Ltd, Essex, England, 1976), p. 279. Egg production is about 12 million annually.

43. MEED, special report, April 1977, p. 14.

44. *Middle East Annual Review, 1978*, p. 299.

45. *Middle East Annual Review, 1977*, p. 279.

5 SOCIAL AND PHYSICAL INFRASTRUCTURAL DEVELOPMENT

Introduction

The pattern of development expenditure evident in Qatar has followed the traditional lines often found in emerging countries in largely emphasising social and physical infrastructure at the early stages of development. Since the country attained independence, all areas of infrastructural activities have expanded in response to the rapid changes brought about by the expanding economy. Despite such expenditure there still are occasional problems, as with the port congestion which can be traced, for the most part, to the post-1973 boom in imports.

The main expenditure heading of the 1972-3 budget included: public works (QR 74 million); Doha sewerage scheme (QR 60 million); Doha port improvements (QR 10 million); Doha International Airport improvements (QR 7 million); compensation deriving from Doha development (QR 40 million); Umm Said industrial estate infrastructure (QR 31 million); road works (QR 51 million); electricity supply (QR 37 million); water supply (QR 33 million); low income housing (QR 13 million); school buildings (QR 9 million); hospital development (QR 6 million); and lastly, television and radio projects (QR 7 million).[1] A careful examination of this list indicates a strong bias in favour of social and physical infrastructure, although it is true that some of the infrastructural projects are part of industrial development schemes, such as at Umm Said.

It is important to note, however, that such a bias in the expenditure structure of the government of Qatar was essential at that stage in order to lay down the vital basis for the development of other sectors of the economy. Indeed, a strong social and physical infrastructural foundation is a prerequisite for the successful implementation of diversification and the push toward balanced growth. The objective is to look at how this foundation has been provided, and to make an evaluation of the increasing role infrastructure will continue to play in the future development.

The Development of Physical Infrastructure

The overall responsibility for major construction projects lies with the Ministry of Public Works, comprised of two sections, the Engineering Services Department and the Mechanical Equipment Department. The former is entrusted with the task of providing civil engineering facilities and liaising with Qatari and non-Qatari contractors and international engineering firms. The latter is charged with the maintenance of various official craft, such as tugs and patrol boats, and operating a 24-hour pilot service involving coastal and harbour navigational beacons and buoys.[2] However, construction forms a significant part of the activities which other expenditures cover.

Transport

Transport problems previously have created serious bottlenecks in Qatar's economy. The average waiting time at Doha's port reached up to as many as 120 days in late 1976; roads were unfinished and airports overcrowded. As a result of the need for a large volume of imports to reach the Qatari economy, total budget allocations for transport in 1976 and 1977 were QR 216 million and QR 563 million,[3] respectively. Consequently, bottlenecks have been lessened and new construction plans have been implemented. The waiting time at Doha port for example, was cut to 10 to 20 days by the end of 1977.[4]

Major development of a modern network of highways and roads began during the early 1950s. Since then, construction has been a process that involves the whole peninsula. There are, at the moment, over 600 miles of main all-weather and rural roads that run throughout Qatar. They link Doha on the east coast and Umm Bab on the west coast, thereby representing the maximum breadth of the country. There are sections in which roadways are not complete, such as Doha to Shamal, Rayyan to Dukkan, Mudeinis to Umm Bab, and roads joining the new airport to the west shore and Umm Said. However, the majority of strategic roadways were expected to be completed by 1978. New additions or maintenance work on currently constructed roads are now designed to allow for future modification into three-lane dual carriage-ways. The Ministry of Transport and Communications plans a coastal roadway to circle the peninsula in such a fashion that all significant villages have close access to the major highway.

In addition to increased internal accessibility through the network of roadways, Qatar has land connections to other countries. A highway, completed in 1969, runs southwest from Doha to the Saudi Arabian border near Salwah and links Qatar with the other Arab states. The road joins the Trans-Arabian system which connects Iraq, Jordan, Syria, and finally Turkey, where it meets the Trans-European road system.

Road transport plays a distinct role in the Qatari economy; it was estimated that in 1975, 35 per cent of Qatar's imports arrived overland,[5] thus relieving some of the congestion at the airport and ports. The 1977 development budget, follow-ing a study by the government, allocated QR 140 million for the continued expansion and maintenance of the roadways.

To relieve congestion and delays at port facilities, several improvements have been made. A major expansion at Doha port was begun in 1976 to add two more berths, and currently the construction of two additional berths is under-way, creating a total of eight. At present there are no plans for adding container-handling facilities, even though they have been needed for years. Only the industrial port at Umm Said will include a container terminal.[6]

Before June 1966, Qatar's only deep-water port was the oil tanker terminal at Umm Said. Due to a wide coral bar situated between Doha's waterfront and the deeper waters of the Gulf, Doha was inaccessible. However, in 1970 Penta Inter-national of Japan was contracted to dredge a channel 27 feet deep, 350 feet wide and 3.5 miles long. When Doha's deep-water port was completed, the estimated total expenditure was QR 144 million.[7]

Since 1971 and the completion of Doha's port, imports have risen steadily. Between 1975 and 1976 cargo distributed through Doha increased by 770,000 tons to 1.2 million tons.[8] At the current rate of development and import requirements, the volume of cargo which Doha port will handle in the future almost certainly will increase. As a result, the present facilities may prove inadequate to support the continually expanding imports. A current proposal is to expand the existing Doha port by constructing a second deep-water berthing quay some 1,000 metres long to service some five ships and thus meet future needs. In addition to the enlargement at Doha port, a similar development is planned at the industrial port at Umm Said, which has a better natural harbour. An estimated amount of $55 million has been allocated for nine wharves, each 200 metres in length. These facilities will be utilised by Qatar Iron and Steel for importing iron ore (two wharves), exporting petrochemicals (three wharves), and the remaining for general freight.

To supplement the expansions, a new port located at Jazirat al-Alyah, 15 kilometres north of Doha, is under consideration by consultants in the United Kingdom. The design includes the building of eight berths during the first stage, with the total potential of the area handling 45 or 50 berths. Since the original proposal in December 1975, the cost of the undertaking has increased 72 per cent, to an estimated $355 million at 1977 prices.[9] Consequently, the Qatari government has reconsidered the necessity for new ports in the future.

Meanwhile, short-run measures have been undertaken to relieve the delays at the existing facilities. One method has been to extend the working time of the dock employees by adding a second shift. This accounted for the reduction of delays at Doha port mentioned earlier and for the reduction in port congestion surcharges, imposed by shippers and insurers, from 85 per cent to 65 per cent.[10]

The present port facilities with their anticipated expansion are perhaps more than adequate to meet Qatar's present and future needs; this fact may be the force behind Qatar's efforts to establish a free zone area in or adjacent to its ports in order to enhance Qatar's entrepôt trading status in the Gulf.[11]

Air transport in Qatar is also being expanded. In 1973, a proposal for a new airport was drawn up, but it has since been abandoned due to the need for a more ambitious project. The need for a new proposal was necessitated by the increase in air traffic and possibly was influenced as well by the prestigious airports being built by Gulf neighbours. Thus, in 1976 the British Airports Authority was requested to design new plans with expanded requirements. The result was a proposal to include 'a terminal with gate facilities for four aircraft up to a BAS 111 in size, four aircraft up to Boeing 707, plus eight aircraft up to Boeing 747 in size, all with nose-in tractor push-out configurations'.[12] The construction of the terminal, which will ultimately handle 900 passengers an hour, is projected to cost QR 340 million at 1977 prices. The new airport, to be located about 17 kilometres north-west of the capital city of Doha, will not be ready for six years, according to a conservative estimate. This location will give the industrial area at Umm Said easy access to the airport facilities.

The Qatari authority's decision to construct a new airport rather than expand the current Doha International Airport is possibly based on two reasons; first, with the extension, pressure on Doha's urban centre would be too great and traffic congestion would result. Second, tall buildings in the capital city could obstruct the flight-path of the planes. In the meantime, the 1976 budget specified QR 25 million for airport expenditures, with the majority of funds going for updating the existing Doha airport; the facility presently handles 30-40 tons of freight a day.

Gulf Air, equally owned by Qatar, Oman, the United Arab Emirates, and Bahrain, is the local airline for servicing the Gulf area. An additional air service, Gulf Helicopters, with headquarters in Qatar, is a privately owned company.

Telecommunications

Telecommunications in Qatar have advanced rapidly over the last twenty years; better communications between Qatar and the rest of the world have been required. Furthermore, good communications among the Gulf states enhance business transactions and increase economic coordination and activities in the region. The 1977 budget set aside QR 45 million for telecommunications development. Approximately two decades ago, Qatar had no modern external telephone and less than five hundred internal communications. In 1953, the exchange system opened with one hundred lines in Doha. The capacity increased to 7,075 lines with the establishment in 1965 of a new exchange in the capital city. Then in 1976, an additional exchange began operation in Doha increasing capacity by 5,000 lines. Other expansions have been made to reach 23,000 sets installed, or more than eleven for every one hundred persons. Most of these units are located around Doha, with the surrounding area connected by land lines and radio circuits. The government has urged this rapid increase in capacity as indicated by QR 45 million allocated in 1976 to telecommunications.[13] The use of the telephone system, regulated by the Ministry of Transport and Communications, is cheap and has helped to encourage increased usage. A single payment covers installation, use of the set, and calls. This subsidy of telephone services in Qatar is yet another avenue through which the government gives discounted services to the population.

Domestic usage is serviced by the Qatar National Telephone Service (QNTS), in which the state has the controlling interest, along with Cable and Wireless (C & W) of the United Kingdom (a 25 per cent holding). In 1976 Qatar's earth satellite communications station constructed by Nippon Electric of Japan for QR 20 million began operation. It links Qatar to all parts of the globe by its thirty international telephone circuits and radio and television communications. The maximum capacity of the station is 5,000 telephone links simultaneously in addition to two colour television channels. Before the satellite went into operation, Qatar used Bahrain's earth satellite station to connect with the rest of the world. Now the facilities enable Qatar to reach Egypt, France, India, Iran, Oman, Pakistan, Saudi Arabia, and the United Arab Emirates by direct dialling. In 1977 direct Kuwait-Qatar and London-Qatar connection began. Except for the government-owned satellite, C & W owns external communications and manages the entire external service. At the moment, the average external telephone time is 500,000 minutes a month. The circuits used for telephones are also utilised by the telex and telegram service. In 1976, a fully automatic exchange opened with 720 lines capable of handling 75,000 outgoing minutes a month. Expansion this year will increase the exchange capacity to 960 lines.

Qatar television, along with other Middle Eastern networks, participates in a 'package' of about seven or eight news items from Visnews, an independent television news agency in the United Kingdom.[14] By access to this satellite broadcasting service, instead of the previously used airfreight method of moving tapes, news events around the globe are transmitted the same day on Qatar television. The fastest airfreight service, in contrast, causes a lag between the events and their broadcast of about twenty-four hours. Visnews blends together a package of about half international news and half events of Arab interest. This is a significant combination since, over the years, Arab broadcasters have traded news items among themselves, under the sponsorship of the Arab States Broadcasting Union (ASBU). However, coverage of items concerning Arab regional interest often are received slowly and frequently lack pictures and quality editorial content. In the field of communications, therefore, Qatar has experienced an accelerated development, compared to other major countries in the Middle East. Moreover, Doha ranks high in terms of telephones per 100 persons, second

Table 5.1: Telephones in Selected Middle East Cities

City and country	Total no. of telephones	Population in 1,000s	Telephones per 100 persons
Abu Dhabi, UAE	14,197	135	10.5
Dubai, UAE	8,041	80	13.6
Ras al-Khaimah. UAE	966	40	2.4
Sharjah, UAE	3,712	50	7.4
Algiers, Algeria	95,132	1,004	9.5
Manama, Bahrain	13,727	109	12.6
Nicosia, Cyprus	37,117	118	31.5
Addis Ababa, Ethiopia	43,923	1,237	3.6
Tehran, Iran	325,200	3,955	8.2
Baghdad, Iraq	102,828	2,872	3.6
Kuwait, Kuwait	128,751	1,031	12.5
Casablanca, Morocco	49,559	1,900	2.6
Rabat, Morocco	34,000	400	8.5
Doha, Qatar	16,195	130	12.5
Tunis, Tunisia	39,704	912	4.4
Istanbul, Turkey	339,659	2,588	13.1

Source: American Telegraph and Telephone Company, *The World's Telephones*, 1976.

only to Nicosia in Cyprus, Dubai in United Arab Emirates, Istanbul in Turkey, and Manama in Bahrain, as Table 5.1 indicates. With developments in the telecommunications network in Qatar after 1975, the country is quickly out-stripping other Middle East countries in this area.

The improved physical infrastructure in transport and communications has concrete implications for the expansion of activities in other economic sectors. For example, the ease with which persons now can reach Doha and the high level of immediate communications available internationally are reflected in new services-linked projects. The Ministry of Public Works raised its budget from $113 million in 1977 to $150 million in 1978 with a major portion of the spending to be devoted for creation of one of the largest conference centres in the Gulf region.

Public Utilities

In Qatar there is a severe scarcity of water, making it expensive. As shown in Table 5.2, precipitation for the area is one of the lowest in Western Asia. In addition to the lack of rainfall, Qatar is deficient in surface and ground level water supplies. Table 5.3 not only brings this fact out clearly, it also indicates that most of the water in Qatar is utilised for agricultural purposes. Ground sources of water constitute the largest supply for Qatar, and most of them are used for irrigation. In 1977, the desalination plants could produce 10.4 million cubic metres of water a year, only about 60 per cent of which was used for domestic and industrial purposes. However, the projection for the year 2000 in Table 5.3 shows that the demand for water for non-agricultural usage will more than triple. However, this need not cause concern for Qatar if the projected desalination capacity of 18.6 million cubic metres annually can be maintained and expansion continued as planned into 2000. One significant aspect which Table 5.3 elucidates is that agriculture cannot depend solely on ground water. By 2000, agriculture will need at least 55 million cubic metres of water per year, compared with the estimated 50 million cubic metres available in 1977. Therefore, given that ground water resources are likely to diminish rather than increase over time owing to the tendency for wells to grow salty, and given that the efforts at drilling new wells are plagued by the law of diminishing returns, the desalination capacity has to be capable of catering to all industrial and domestic uses and, in addition, supplementing the ground water supply to keep agriculture going. Water management and expanding alternative sources of water supply become essential for continued development in the industrial, agricultural and other sectors of the Qatari economy and to support the growing population.

There are several methods through which an increase in Qatar's water supply may be achieved; most of these means are expensive. Desalination is one of the methods which has been adopted by most Gulf states. Kuwait has succeeded almost completely in using desalination as a means of supplying its water needs. There are, of course, more apparently exotic alternatives, such as towing icebergs from the Antarctic region, an option being considered in Saudi Arabia and even in California.[15]

Qatar, like Kuwait, uses the multi-stage flash (MSF) evaporation technique in desalination. This method operates by heating large amounts of sea water at various stages while it is flowing inside copper alloy tubes. Finally, in the last stage, the saline solution is heated by steam and then routed into pressurised flash chambers. The pressure is reduced slowly to permit evaporation, while vapours condense on the outside of the tubes producing the condensate which becomes the desalted water. The desalination process is very expensive, requiring vast amounts of equipment and fuel. The steam used to promote evaporation comes from an external supply. As a result, many installations arc located near power plants to utilise their exhaust. In any case, the heating process requires that the temperatures be carefully controlled. Equipment must constantly be replaced, as the copper alloy tubes often corrode. It is estimated that the MSF process, or any other method currently feasible, costs $ 1 for every cubic metre of desalted water — with energy at $ 80 a ton.[16]

The water department in Qatar was established in 1954 and almost immediately began installing desalination units. That year, supervision of three submerged coil units started, along with their own steam-raising unit on a site adjoining the original capital city diesel power plant. This plant is

Table 5.2: Average Annual Precipitation in Western Asia (millimetres)

Bahrain	75
Iraq	
Desert plateau	50-150
Lower Mesopotamia	100-200
Upper plains and foothills	300-500
Mountains	600-1000
Jordan	
West highland	300-700
East highland	300-600
Jordan valley—Dead Sea	—
Southern desert	below 250
Eastern desert	below 150
Kuwait	120
Lebanon	
Coast	800
Beqaa valley	250-1500
Mountains	900-1500
Oman	40-180
Qatar (Doha)	60
Saudi Arabia	
North region	80-120
Northeastern region	50-70
Centre	85-110
Red Sea coast	250
Mountains	400
Syria	
Coast and mountains	500-1000
Other regions	50-500
UAE	65
Yemen (Aden)	
Mountains	above 400
Coast	above 50
Northeast	almost nil
Yemen (Sanaa)	
Plains	200
Mountains	800

Source: MEED, 29 April 1977, p. 20.

Table 5.3: Water Resources and Needs (million cubic metres a year)

	Water resource potentials					Water use			Future water demand			
	Surface ground[a]		Desalinated									
			Present	Future	Year of completion	Agriculture	Domestic	Industry	Agriculture	Domestic	Industry	Year of projection
Bahrain	Negligible	199	8.3	24.7	(1981)	166	20	13	126[b]	41[c]	15	n.a.
Iraq	80,000	n.a.	n.a.	n.a.		39,530	580	2,240	52,100	3,520	11,960	(1995)
Jordan	850	165	n.a.	n.a.		375	4	6	465	60	30	(1990)
Kuwait	Negligible	130	102.9	66.4	(1980)	130[d]	75	8	1,150	1,730	50	(1995)
Lebanon	3,800	50	n.a.	n.a.		647	94.0		3,180	600		(2000)
Oman	10	665	2.0	n.a.		420	10.0		n.a.	n.a.		n.a.
Qatar	Negligible	50	10.4	18.6	(1980)	44	6.0		55	20		(2000)
Saudi Arabia	2,200	1,725	17.8	128.8	(1977)	13,500	830	150	32,400	250	1,048	(1980)
Syria	32,000	1,600	—	—		6,000	400	n.a.	18,000	1,500	n.a.	(1990)
UAE	160-270	270	2.0	n.a.		331	31.3		409	42		(1990)
Yemen (Aden)	1,500	350	n.a.	n.a.		1,900	26.0	n.a.	n.a.	n.a.		n.a.
Yemen (Sanaa)	n.a.	n.a.	n.a.	n.a.		n.a.	n.a.	n.a.	n.a.	n.a.		n.a.

a Bahrain, UAE: safe yield; Oman: exploitable amount; others: production.
b 1980.
c 1985.
d Including garden and household use.

Source: MEED, 29 April 1977, p. 14.

referred to as the Number 1 Distillation Plant. The following year witnessed the extension of the installation by an addition of two duplicate units to produce a total of 130,000 gallons per day (g/d) of desalted water. In 1959, the Number 2 Distillation Plant was completed, consisting of 150,000 gallons per day flash units. Four years later, the Ras Abu Aboud complex was built with two MSF units, each having a designed output of 750,000 gallons per day. The years 1967 and 1968 saw the implementation of the plans for Phase II expansion of the Ras Abu Aboud complex with the installation of two more units to contribute two million gallons per day to the water supply. By mid-1969, the implementation of a three-phase expansion programme was initiated in an attempt to supply Doha and the surrounding area with a safe, fresh water supply.[17]

Due to the rapid economic growth Qatar has experienced in the 1970s, water consumption has risen at an alarming rate. In December 1972, total consumption was 6 million gallons a day compared to 10-10.5 million in 1975. Natural water sources are able to supply only 3.2 million gallons daily. Unfortunately, as underground water is utilised, the aquifer sinks and the fresh water becomes salty. Consequently, the government has adopted a plan with a 1979 target of dependence entirely on alternative sources, which essentially means desalination.

Since the Ras Abu Aboud installation, new plants have been proposed and completed to obtain the desired capacity of 32 million gallons per day. One such plant is the Ras Abu Fontas electric power and desalination complex on the Gulf coast, south of Doha. The first phase of the project was completed in 1977 at a cost of QR 470 million ($ 199 million), consisting of two gas turbines with a capacity of 100 megawatts and two desalination plants. The total undertaking, scheduled for completion in 1979, is estimated to cost QR 1,949 million ($ 493.3 million) and have a productive capacity of 32 million gallons a day of desalted water and 612 megawatts of power. The second stage, to be completed in 1978, at a cost of over QR 429 million ($ 108.6 million), will include four additional desalination installations with four gas turbines to operate at a capacity of 224 megawatts. The final stage with a price tag of QR 1,050 million ($ 265.7

million) will increase electrical output by 288 megawatts as a result of the construction of six extra gas turbines and two more desalination plants.

Construction of the second phase of the power station with four gas turbines was awarded to a West German firm, Kraftwerke Union (KWU) at $68.7 million. Societa Italina Resine (SIR) received a $38 million contract for the desalination plant. KWU was previously contracted to construct two gas turbines for the first stage of the complex and SIR provided the desalination plants. In addition, Fiat Termoneccanica Nuclearee Turbogas of Italy, a subsidiary of the Turin-based Fiat group, was awarded an $8.5 million contract for construction of a power station. Other major desalination plants soon to operate include three new units at Ras Abu Aboud (3 million gallons a day), Wakrah (4 million gallons a day), and Umm Said (12 million gallons a day).

Around Doha, major recycling has been introduced to provide the capital city with its fresh water supply. Doha at present has a modern sewage system in which water is treated, stored in towers, and then recycled. Through the construction of desalination and power plants, the Ministry of Electricity and Water is attempting to increase electrical power output at a pace to match the expanding economy's demands. In 1975 the country's electrical power capacity was only 100 megawatts; by 1978 it was expected to grow to at least 500 megawatts. To reach this generation of electrical power, the Qatari Development Budget in 1976 designated QR 399 million and QR 185 million to electricity projects and water supply schemes, respectively. In 1977, the budget for electricity was increased to QR 908 million.[18]

There is no doubt that for centuries fresh water has been a precious commodity in the Gulf region. In Qatar, as with all arid regions of the world, adequate supplies of fresh water cannot be taken for granted if industrial, agricultural, and urbanisation programmes are to be implemented smoothly. As the economy grows, the demand for fresh water for domestic, industrial and agricultural purposes will continue to increase. In recognition of this, the Qatari government has given top priority to ensure that a sufficient supply of fresh water is available, as evidenced by the many desalination projects either undertaken or proposed by the government and its agencies.

A possible alternative, which may sooner or later be considered, is what may be code-named a 'Saudi dream'. The dream revolves about the towing of icebergs from Antarctica to be used to supply fresh water in Saudi Arabia. It is estimated that if it is successful, the iceberg project could cut the production cost for a cubic metre of fresh water by about 33 to 73 per cent. Although the location of Qatar may make it more expensive to use icebergs than it would be in neighbouring Saudi Arabia, if the Saudi dream becomes an economic reality, Antarctic icebergs could become alternative sources of fresh water supply. To this end, a joint venture with Saudi Arabia, and/or other Gulf states in which Qatar participates, may prove beneficial.

Social Development

It has long been recognised by social scientists and by economists that a country's development does not stop with the growth of economic well-being. Thus, development is recognised as having a broader connotation, which includes the expansion of economic and social amenities, as well as improved distribution of opportunities. Total development, then, consists of anything which contributes to the process of making the average citizen experience an increase in his welfare.[19]

In capital-surplus countries, the scarcity of investment opportunities in purely economic sectors has made social infrastructure a more attractive sector in which the government spends accumulated oil revenues. In these states, including Qatar, the wider concept of development or welfare is more closely approximated than in countries with large investment opportunities in the various economic sectors. Qatar may be regarded as falling into the high level of welfare state stratum. Educational and health services are provided free of charge to the population; in addition, the government spends large sums of money on the provision of other social and welfare services.[20] Let us turn to examine in some detail the socio-cultural aspect of Qatari development.

Education

Education, along with other social development sectors in

Qatar, has moved forward briskly. Table 5.4 gives evidence of the pace of change. In 1952, the first boys' school was established with only 240 students. Pupil numbers have increased almost a hundredfold to 20,531 in 1972. At the same time the number of teachers rose from 6 to 1,103 and the number of schools expanded to 85 in 1972. In 1955, the first school for girls began operation with only 50 students. In spite of some initial cultural-based opposition, the percentage of girls to total school population rose consistently to constitute about 42 per cent in 1971. In 1977, total school population has been put at 31,000, of which approximately 50 per cent are female. In 1978 the government plans to open 22 new schools.[21]

The education structure is divided into primary (six years), preparatory (three years), and secondary (three years) stages, commencing at the age of six. Primary education is the only compulsory level for boys and girls; it is expected that eventually the preparatory and secondary stages will be compulsory. To achieve this goal, the Ministry of Education has plans to construct 24 elementary and 14 secondary schools by 1982. There are also four specialised secondary schools: the Doha Technical School opened in 1956, and the Secondary Commercial School, the Teacher Training Institute, and the Institute of Religious Studies all began classes in 1966.

A new development is the introduction of the kindergarten. By 1972, there were 16 kindergartens with an enrolment of 2,595 pre-school children. Although it is too early to have reliable survey findings, the availability of these institutions may encourage female participation in the labour force, and, as noted earlier, there is an insufficient indigenous labour supply.

Higher education is as yet unavailable in Qatar, but on 22 February 1973 the Emir declared that the government had 'laid down the bases for the establishment of a Teacher's College as the first step towards the envisaged Qatar University . . . which we shall make available to all citizens of the Gulf region'.[22] As a result, in 1976 the women's and men's Teacher Training Institutes were joined together in an attempt to establish a University of the Lower Gulf. A budget of QR 160 million has been established for the university campus, scheduled to open in 1982 with an estimated enrolment of

Table 5.4: The Development of Education in Qatar, 1951/2–1971/2

Academic year	Number of students			Number of teachers			Number of schools		
	Boys	Girls	Total	Men	Women	Total	Boys	Girls	Total
1951/2	240	—	240	6	—	6	1	—	1
1952/3	320	—	320	10	—	10	2	—	2
1953/4	457	—	457	16	—	16	2	—	2
1954/5	560	—	560	26	—	26	4	—	4
1955/6	1,000	50	1,050	45	1	46	15	1	16
1956/7	1,388	122	1,510	50	4	54	17	1	18
1957/8	1,879	452	2,331	105	14	119	22	2	24
1958/9	2,437	578	3,016	163	26	189	25	5	30
1959/60	3,235	1,423	4,658	286	91	377	27	11	38
1960/1	4,023	1,942	5,965	359	135	494	40	20	60
1961/2	4,610	2,450	7,060	376	144	520	46	21	67
1962/3	5,364	2,715	8,079	410	175	585	51	24	75
1963/4	6,145	3,381	9,526	428	205	633	50	27	77
1964/5	6,981	3,872	10,853	455	214	669	50	28	78
1965/6	7,912	4,811	12,723	537	254	791	50	31	81
1966/7	8,301	5,405	13,706	553	303	856	50	33	83
1967/8	8,875	5,651	14,336	596	325	921	50	35	85
1968/9	9,371	6,281	15,652	547	352	899	47	37	84
1969/70	10,122	7,101	17,223	n.a.	n.a.	n.a.	n.a.	n.a.	n.a.
1970/1	10,704	7,827	17,531	n.a.	n.a.	n.a.	n.a.	n.a.	n.a.
1971/2	11,608	8,923	20,531	573	530	1,103	44	41	85

Sources: *Malif al-Nahar*, vol. 32, 1969, p. 44 (in Arabic); *Qatar into the Seventies*, pp. 67-8.

2,000. Until then, approximately 600 Qataris are studying abroad to receive their higher education.[23] Towards the end of 1977, a planning committee was nearing the end of its work on the establishment of a School of Engineering within the University. The advisory group consisted of engineering professors from such American Universities as Oklahoma, Iowa, and Louisiana. UNESOB experts were also consulted on the project. The new College of Engineering should provide a major advance in training the Qatari population, particularly in fields relevant to industry and to physical infrastructure programmes.

Two institutions for in-service training deserve special notice: the Management Institute and the Regional Center for Vocational Training. The first was initiated in 1970. It offers a two-year course on subjects ranging from administration to accounting, public finance, public relations, economics, statistics, international law, correspondence, and English. Senior officials with a secondary education are admitted provided they have reached at least the intermediate cycle of government-grade classification.[24] The other in-service training institute is the Regional Center for Vocational Training, started late in 1971 with the assistance of the United Nations. The essential aims of the project are: (a) to establish national policies and short- and long-term vocational training programmes; (b) to provide technical and functional training for instructors and shopfloor supervisors; (c) to organise apprentice training programmes; (d) to upgrade and improve skills of artisans; and, (e) to organise accelerated training courses for adults.[25] The Center has international experts and expatriate instructors in addition to Qatari instructors. The regional character of the Center is reflected in its student body that comes from five of the Gulf sheikhdoms which partially comprise the United Arab Emirates — Sharjah, Ajman, Umm al-Qawain, Ras al-Khaimah, and Fujairah. The Center was established with eleven training sections: auto-mechanics, machine shop, welding, carpentry, building, plumbing, electrical, refrigeration and air conditioning, radio/television, drafting/surveying, and clerical.[26]

The second phase of the Regional Center for Vocational Training opened in early 1973; it aims at consolidating the Center's achievements to date and expansion by means of a

30-month plan under the direction of the government in co-operation with the United Nations Development Programme (UNDP). The training of instructors and supervisors along with the settling of trade standards, testing, and certification levels was emphasised. Seven international experts for a total service of 178 man-months, 42 man-months of fellowship training of Qatari personnel, and training equipment were supplied by the United Nations Development Programme. As part of the programme's second phase, the Qatar government and UNDP contributed $ 1,805,600 and $ 530,900, respectively.[27]

The Qatari government has shown special concern for progress in the field of social services and education since these have a direct impact on the development of Qatari human resources. Substantial expenditures are involved; the 1977 budget earmarked QR 895 million for education. All aspects of education have been considered. In addition to kindergarten, compulsory primary education for boys and girls, technical schools and in-service training centres, the Ministry of Education has supported adult education. School buildings left vacant in the evenings are used for intensive instruction to combat illiteracy and to prepare the participants for the primary school certificate. Not only are these services available, they are provided by the government free of charge at all levels:

> The government is wholly responsible for tuition and maintenance up to and including higher education outside the country. Books, school meals, transportation, clothing, holidays and boarding accommodations are provided for. Monthly cash allowances are made available to needy parents and pupils. Overseas scholarships are generously awarded; the successful candidate can confidently relish the prospect of travel in the knowledge that this will not entail family sacrifice and that his pocket money and living allowances will be related to the foreign cost of living.[28]

The development programme for education has not been without obstacles. Of the 24 elementary and 14 secondary schools scheduled for completion in 1982, only three were operational by the end of 1976. The general phenomenon of the dominance of the non-Qatari in the population mix, labour force, and the like is similarly encountered in the field of education. The result is that the majority of students are non-Qatari. Therefore, it cannot be assumed that all the graduates and trainees will remain in Qatar permanently, with obvious implications for manpower projection and planning. The teaching profession is also dominated by non-Qataris. In 1970-1, there was a total of 162 Qatari teachers at the primary school level (177 males and 45 females); this contrasts with 590 non-Qatari teachers (290 males and 300 females). At the secondary level, there were only 10 Qatari teachers, while expatriate instructors numbered 311.[29] However, this trend is undergoing a reversal, albeit a slow one, given the very time-consuming nature of the process of education, as Qataris are trained to assume teaching positions.

Health

Medical, dental, and surgical services are provided free to all — Qataris and expatriates alike. Home visits by specialists and physicians, hospitalisation, provision of drugs and special treatment, and even care in the United Kingdom, Austria, Lebanon, or wherever needed, are included in this social service.

Substandard health conditions prevailed before the advent of oil revenues. Harsh climatic conditions along with extreme poverty contributed to the high incidence of a variety of diseases, such as malaria, tuberculosis, dysentery, and trachoma. Until oil funds accumulated, Qatar lacked all types of medical staff and health facilities as well as the financial resources necessary for their importation and maintenance. The progress in recent years is all the more remarkable if it is remembered that as short a time ago as 1945, the sum total of the country's official medical resources amounted to a single clinic with a single resident physician.[30]

As the 1970s began, the Ministry of Public Health proposed the improvements needed for health service development based on a survey it had conducted. At that time, there were three modern government hospitals and two private hospitals with an overall total of 658 beds. During this same period, Qatar had 64 physicians, 5 dentists, 16 pharmacists, 160 nurses, and 31 orderlies.[31] Additionally, five clinics served the population. Normally the patient visits the clinic first, and if it is deemed necessary, he is then admitted to the hospital. As a result of the Ministry of Public Health's survey and of recommendations from consultants in the United Kingdom, a 20-year health programme was formulated. This was an all-embracing plan developed to cover several aspects of health care. The plan considered medical reform, providing health care and treatment, clinics, pharmacies and laboratories, construction and maintenance of hospitals, and even research associated with international organisations with regard to endemic diseases.

At present, the plan includes completion, in 1978, of one of the most advanced hospitals in the Middle East in Doha; it will possess some 650 beds and complement the existing city hospital at Rumaillah. QR 160 million has been allocated for this new modern amenity which will include intensive therapy and care units, a paediatric unit, six operating theatres plus one for complex surgery, and the Ministry of Public Health's administrative offices. For future additions, facilities for major diagnostic nuclear medicine and physiotherapy are planned. Llewelyn-Davies, Weeks, Forestier-Walker and Bar of the United Kingdom, specialists in hospital design, were commissioned to plan the hospital. The actual construction is being done by another British firm, Bernard Sunley and Sons. At the same time, an addition to the existing Rumaillah hospital is planned that will increase its capacity by 64 beds and provide for chest, geriatric, and psychiatric cases. The fact that the new extension is to include facilities for psychiatric cases is an achievement, for a developing country does not usually acknowledge psychiatric ailments. Included in the expansion are increased maternity and specialist child care features.[32] The Ministry of Public Health, in attempting to expand the medical services which had consisted only of primary hospitals associated with clinics, has proposed the construction of 10 health centres in the interior to serve 20,000 people.

Social Welfare

In addition to the provision of free education at all levels, free medical services, and the subsidising of public utility rates to non-industrial as well as industrial users, the Ministry of Labour and Social Affairs administers the government's 'popular' housing, home ownership, and social welfare schemes. Housing programmes are critical and respond to the very real needs of many Qataris, as rents and land prices are extremely high. Acknowledging this problem, the Emir, marking his accession to power, announced in February 1972 that 659 housing units would be distributed free of charge and that outstanding repayments on 1,892 more would be cancelled. The government's continued concern for housing programmes is evident by the 1976 budget which allocated QR 50 million for 'popular' housing and set aside QR 140 million to be used for providing relatively interest-free loans to the private sector to build dwellings. The 1977 and 1978 budgets allocated 10.2 and 17.8 per cent, respectively, of the development spending to housing.

Public housing is available to low-income families; the government provides the recipient with land and/or a house. There is flexibility in the programme by allowing the grantee who has previously acquired land to build a structure with government financial support. Only 70 per cent of the overall cost must be repaid to the government if the low, interest-free instalments are regularly met over a period of 20 to 25 years. Assistance is also rendered with furnishings within the repayment arrangements. Should the beneficiary die, the government automatically relinquishes its rights.[33] Thus far, a total of 1,500 low-cost houses have been constructed, and the 1977 budget includes funds of QR 80 million for 600 more. The 1978 budget earmarked over $ 100 million for public housing.[34]

In regard to private sector housing development, the government has recently made loans available to potential builders and buyers. For example, up to QR 30,000 may be given as a loan for construction of a private home and a grant of up to QR 25,000 may also be obtained for furnishings.[35] The private sector has to date (1977) not been very responsive to this programme to stimulate an expansion in housing. This has been attributed to the fact that larger profits can be realised through land speculation as well as to the shortage of skilled construction workers and the necessary supplies. As a result of these shortages, the quality of the construction can be endangered. In the long run, this could result in the country's capital stock being depreciated at a faster rate than necessary, thus causing increased future expenditures when Qatar's revenue may be limited.

In addition, after investigation and evaluation by qualified social workers, the government administers regular monthly cash allowances scaled to individual requirements of orphans, widows and others in need.[36]

The upgrading and better utilisation of human resources through education, health, and other social services go hand in hand with physical infrastructure expansion, and both are necessary inputs to the economic development process. Of the two the physical can be more speedily accomplished; roads are built more quickly than a person can be trained as an engineer, physician, or accountant. In the case of Qatar (and in common with the other oil-based developing countries), rapid economic growth initially is most easily and logically channelled into physical infrastructure. The appearance of first-class roadways where few existed before, thoroughly modern airports, telecommunications, ports and marine transport-related docks and facilities can very quickly transform the face of a nation. Moreover, these basic endeavours and projects are prerequisites for industrial, agricultural, and service-sector expansion. Such investment in physical infrastructure can be made by importing required skills and labour.

But perhaps the key to the formula for translating economic growth into self-sustaining development lies in the less obvious, longer-term matter of investment in human capital. Better health and education increase the population's productivity. It is an expenditure with a lag between the outlay of funds and the beginning of returns. But without such human infrastructure, the bottlenecks to the development process will remain. Qatar, as with the developing bloc of countries in general, is involved not only in training engineers and economists, but it must eventually have a full range of highly skilled, skilled, and less-skilled workers to fill a wide spectrum of managerial, technical, and labouring slots. Thus, the admittedly advanced state of the social services sector should be evaluated in this light.

Finally, physical and social infrastructure development has served as a relatively simple and rather immediate means of income distribution. Since the oil revenues accrue directly to the government, spending on roads, ports, and the like, means jobs and opportunities. Government-financed education, health, low-income housing, and welfare are all a further way of spreading the oil-based national wealth among the population.

Notes

1. Ministry of Information, Government of the State of Qatar, *Qatar into the Seventies* (Ali bin Ali Printing Press, Doha, 1973), p. 106 (hereafter cited as *Qatar into the Seventies*).

2. *Qatar into the Seventies*, p. 89.

3. See Table 4 of the chapter on public finance and economic policy.

4. MEED, 28 October 1977, p. 32.

5. *Middle East Annual Review, 1977* (The Middle East Review Company Ltd, Essex, England, 1976), p. 273.

6. MEED, 29 April 1977, p. 39.

7. For a full description of the port and related matters, see *Qatar into the Seventies*, p. 93.

8. MEED, 29 April 1977.

9. Ibid., p. 16.

10. MEED, special report, April 1977, p. 15; see also p. 96 of this chapter.

11. *Qatar into the Seventies*, p. 91.

12. *Mideast Markets*, 14 March 1977, p. 7.

13. MEED, special report, April 1977, has an informative section on the improvements of telecommunications in Qatar, pp. 16-17.

14. MEED, 27 May 1977, p. xv.

15. MEED, 9 September 1977, p. 31. Also see a later discussion of this novel idea.

16. MEED, special report, April 1977, p. 15.

17. The summary of activities from 1954 to 1969 is covered in more detail in *Qatar into the Seventies*, pp. 82-3.

18. See the chapter on public finance and economic policy for development budget allocations for 1977 and 1978.

19. James Tobin and William D. Nordhaus have advocated a new method for measuring welfare other than the use of gross national product (GNP) concepts. The new concept they put forward is the Measure of Economic Welfare (MEW) which takes account of values of leisure, non-market activities, and disamenities such as urban pollution, etc. James Tobin and William D. Nordhaus, *Economic Growth* (National Bureau of Economic Research (NBER), New York, 1972; NBER Colloquia Series on Economic Research: Retrospect and Prospect, no. v).

20. See chapter on public finance and economic policy for an over-all view of government finance.

21. MEED, special report, April 1977, p. 9; and *Mideast Markets*, 10 April 1978, p. 2.

22. United Nations Economic and Social Office in Beirut (UNESOB), *United Nations Inter-Disciplinary Reconnaissance Mission to Qatar* (July 1972), p. 69.

23. *Qatar News*, July-August, 1977, p. 1; MEED, special report, April 1977, p. 20.

24. UNESOB, pp. 70-1. The Management Institute's enrolment in 1972 was eighteen. The Diploma of Administration is awarded to participants who complete the required course.

25. Ibid., p. 70.

26. *Qatar into the Seventies*, p. 72.

27. Ibid., pp. 72-3.

28. Ibid.

29. *Pakistan Monitor*, 22 February 1973, p. 66.

30. Muhammad T. Sadik and William P. Snavely, *Bahrain, Qatar, and the United Arab Emirates* (D.C. Heath and Company, Lexington, Mass., 1972), pp. 99 ff.

31. Ibid., p. 103 f.

32. MEED, special report, April 1977, p. 19.

33. *Qatar into the Seventies*, p. 74.

34. MEED, special report, April 1977, p. 18; *Mideast Markets*, 10 April 1978, p. 3.

35. Ibid.

36. *Pakistan Monitor*, 22 February 1977, p. 27.

6 MONEY AND BANKING

Introduction

Although the debate about the influence of monetary aggregates on production, employment, and economic growth has not been settled to everyone's satisfaction, all parties at least agree that the monetary and real sectors of an economy are invariably interconnected. In nations the economic activities of which are built upon the revenues of a single commodity, such as oil, and in which all the revenues accrue to the government, the need to disburse the huge revenues renders the entire economy dependent on governmental budgetary policies. As a result, both monetary and fiscal policies determine the level of production, consumption, and investment in such economies. As is the case in most surplus-funds countries, the monetary and fiscal policies of Qatar are so closely related that they can be taken as one and the same thing. For the flow of money into the economic system depends on the extent to which the Qatar government spends from its oil revenues.

Qatar attained independence only relatively recently, yet the financial development in this part of the Gulf has a long history. A short review of these developments is necessary for a better understanding of the present monetary and financial arrangements in Qatar, especially those political and economic factors which precipitated the changes in the institutional structure of the money and banking system. Based on these developments, the present money and banking system, the monetary and fiscal policies and their implications for economic development are analysed in the rest of the chapter.

Early Financial Development

The Pre-Currency Authority Era

The early development of banking and financial institutions in Qatar may be discussed more effectively within a regional context, i.e., within the framework of currency developments

119

in the Arabian peninsula.[1] This is a convenient approach not only because of the difficulty in obtaining information on financial developments pertaining to Qatar alone, but also because during this early period the flow of trade and capital in this region was so unconstrained that it is more valid to treat the whole peninsula area as the unit of analysis.

The history of money in the Arabian peninsula followed the formula found in other parts of the world. After experimenting with several objects as a means of exchange, metallic coins came into use as money. Owing to the early trade with East Africa, India, and other Asian countries, it is possible that the period in which non-metallic money circulated in this region before fully fledged coins were introduced might have been shorter than has been found for most parts of the world. The most widely circulated of these coins were the Maria Theresa thaler and the Indian rupee; the former was first minted in Austria in 1780.

The two coins were introduced into the region through international trade. It appears that the Indian silver rupee was the dominant currency circulating in Qatar prior to the introduction of paper money; the Maria Theresa thaler was more widely used as a medium of exchange along the southern coast of the Arabian peninsula.

The traditional disadvantages inherent in a metallic currency system caused the abandonment of the system and led to the extensive use of paper money throughout the Arabian peninsula, with the exception of Saudi Arabia which continued to use only coins into the second half of this century. Thus in Qatar, the difficulty in transporting large amounts of coins, their use for large transactions, and the full-bodied nature of the coins which made them an unstable monetary unit, all combined to reduce the use of coins and the emergence of the Indian paper rupee as the principal currency. This continued until May 1959 when the Indian government, in an attempt to stop black-marketing in the currency, issued a new rupee to be used in the Gulf states. The value of the Gulf rupee, as it was named, was at par with the Indian rupee (0.186621 grams of fine gold).

Further currency developments in the Gulf area consisted of the issue of national currencies, the establishment of central monetary authorities, and currency boards. Qatar was among those countries which did not jump from a Gulf-rupee standard directly to a Qatar national currency system. Instead, a currency authority was established for Qatar and Dubai. These and future financial developments took place against the background of rapid political and economic changes.

As mentioned earlier, Qatar achieved full independence in 1971, together with other members of the so-called Trucial states. There was a natural and understandable desire among these new nations for their own monetary systems. Whereas this political event influenced the establishment of a monetary agency perhaps more than the decision to have a currency authority with Dubai, the economic changes had a more compelling influence on both stages in the move towards an independent monetary system in Qatar. The discovery of oil in Qatar in the 1930s gradually but consistently increased the rate of monetisation of the Qatari economy.

The Currency Board Period

With the growing oil wealth came the need for a Qatari national currency. Following the formula which Britain had used widely in its overseas territories and colonies, the 'colonial exchange system' was introduced in most of the Arabian peninsula countries which were under the protection of Britain.

In 1966 a currency agreement was signed between Qatar and Dubai to establish the Qatar-Dubai Currency Authority (QDCA). Like all currency boards, its functions were limited to the issue and redemption of currency which was backed at least 100 per cent by sterling reserves of the two countries. This made the currency issued by the QDCA a 'sterling surrogate'.[2] As if attempting to take on a characteristic of a central bank (in addition to those of currency issue and redemption), the currency authority restricted its operations to commercial banks and the government. It is doubtful, however, if the QDCA was able to perform the function of banker to the government and commercial banks; the commercial banks, for example, only served as intermediaries between the Authority and the private sector. The Qatar-Dubai Currency Authority was therefore essentially an

automatic money changer with no powers regarding monetary policy and general control over the financial/banking
system in either Qatar or Dubai.

The relationship between the Qatari money market (despite
its underdeveloped nature) and the London money market
in this period transcended that aspect brought about directly
by the activities of the currency authority. Indeed, with the
currency authority powerless in influencing liquidity in the
country, flexibility in the Qatari financial system was introduced only by the local commercial banks, which were
branches of banks in the metropolitan country.[3]

After the establishment of the Authority by the signing of
a currency agreement between Qatar and Dubai on 21 March
1966, the issuing of the Qatar/Dubai riyal was not immediate.
The devaluation of the Gulf rupee, to which the riyal was
tied, in June 1966 caused the introduction of the new
currency to be postponed until 18 September 1966. In the
six-month interim, the QDCA borrowed 100 million Saudi
Arabian riyals for circulation in the two member countries,
in place of the Gulf rupee. The issue of the Qatar/Dubai
riyal valued at 0.186621 grams of fine gold (i.e., at par with
the pre-devaluation Gulf rupee) was important not only for
Qatar and Dubai but also for the other Trucial states —
Bahrain, Oman, and the sheikhdoms comprising the United
Arab Emirates with the exception of Abu Dhabi — where
the Qatar/Dubai riyal became the sole legal tender currency
for some time.[4]

Commercial banking in Qatar during the pre-Central
Monetary Agency (CMA) era was dominated by British-
owned commercial banks. The main concentration of operations of these banks has been in export and import trade,
with their lending and investment policies determined by
their head offices in Britain. Therefore, even in the case of
demand or time-saving deposits, no central monetary
institution in Qatar wielded control. Coupled with the passive
activities of the Qatar-Dubai Currency Authority, there was
virtually no monetary policy.

The 'colonial exchange standard' has been criticised for
tying the money supply and the monetary policy of the
territories concerned to those of the British economy.[5] The
Qatar-Dubai Currency System may be defended on the

grounds that it was a better system than that which had existed in Qatar. For the first time, the issue of currency was centralised, yet the amount of money which circulated in the nation continued to be determined by the balance of payments of the country to the extent that 100 per cent reserve banking was required. Even now with the Qatar Monetary Agency, it cannot be said that monetary policy in Qatar is completely independent of its balance of payments, although, as will be seen later, fiscal policies may be tailored in such a way that some of the balance-of-payments effects are neutralised. In any case, the currency board system served as a stepping-stone to the establishment of a full-fledged central banking system.[6]

Further, it is believed that control of the credit activities of commercial banks in Qatar during the currency board period was not necessary so long as banking practices of the British commercial banks were not contrary to the interest of Qatar.[7] While this may be a difficult notion to entertain, perhaps it was not until other foreign banks entered the picture by opening branches in Qatar that the emergence of divergent banking practices necessitated the establishment of a central monetary agency charged with the duty of providing a unifying direction to the banking system.

The Central Monetary Authority Era

The currency board system became too restrictive. In Qatar the oil boom in the 1970s made the establishment of a separate central monetary authority imperative. As will be seen later, the instruments needed for implementation of monetary policy have to date not been fully developed by the Qatar Monetary Agency (QMA). In any case, on 19 May 1973, the QMA was established by a royal decree.

The purpose of establishing the Agency was clearly stated in the 1973 law. The QMA is responsible for regulating, redeeming and supplying currency, the Qatar riyal. It was also charged with the duty of ensuring both the internal and external stability of the riyal and safeguarding its convertibility into other currencies. Another objective is to 'promote credit and exchange conditions conducive to the growth of the economy within the framework of monetary stability', and lastly, the QMA has a supervisory power over the banks in Qatar.[8]

It will be noted that these functions of the QMA do not include explicitly those of serving as the banker to the government, nor as the bankers' bank. Nonetheless, it does not follow that the functions could not be performed. Indeed, the functions of safeguarding the external and internal value of the riyal, ensuring its convertibility, and promoting credit and exchange conditions favourable to development could be performed efficiently only if the commercial banks' credit operations could be controlled and if the Agency holds government credit instruments. The broad manner in which the functions of the Agency were specified in the 1973 law is an indication of the flexibility given to the monetary authority in its attempts to exert control over the Qatar monetary system.

It is not only the development-oriented nature of the functions of the QMA which makes it a member of the emerging group of central banks for developing nations. Its close relationship with and almost complete control by the government of Qatar constitute a break with the old version of central banking legislations. The QMA is completely owned by the government of Qatar, and all the seven members of the Board of Directors are appointed by the Emir, upon the recommendation of the Minister of Finance. To this extent, the independence of a central bank which, in Western monetary arrangements (especially that of the United States), is considered a vital element in the performance of stabilisation functions, does not exist in the Qatari case. However, the government invariably takes into consideration the recommendations of the Monetary Agency, and since monetary and fiscal policies are interlinked, the relationship between the government's financial wing and the QMA should necessarily be cordial. Such a relationship may be ensured, to a large measure, by the interdependence of the two bodies, as reflected in the appointments of the Board of Directors of the Agency and the interest that the Minister of Finance has in other activities of the Agency. These involvements of the government in the activities of the QMA consist of approvals which the Minister of Finance must give to the Agency's recommendations regarding changes in the external value of the riyal and the external reserve backing of the currency.

The Central Monetary Agency in Qatar does not have much control over the banking/monetary system. Neither does it have a monopoly over the government agency activities in financial transactions. In other words, apart from issuing of currency and ensuring 100 per cent foreign exchange backing of the riyal, the field of operation of the QMA is restricted. In Qatar, the central bank does not carry on commercial banking, but the reverse is true. The Qatar National Bank (QNB), a commercial bank, performs central banking functions. The QNB, half-owned by the government, is the principal channel through which government funds are disbursed and, instead of the Qatar Monetary Agency, is the pivot of the Qatar financial system.

The reasons for this state of affairs in Qatar's financial market are not difficult to identify. In the first instance, the QMA was established only five years ago, and as a result may still be in what might be termed the infant stage so far as control over the monetary/financial system is concerned. Second, the Agency was established in a period of accelerated increase in financial wealth in Qatar, and thus, despite the store of experience inherited from its predecessor, the Qatar-Dubai Currency Authority, it might have been a deliberate policy to allow the more experienced QNB to continue to be the central focus of the financial system. This would enable the QMA gradually to take over control when it finds itself capable of doing so. The tendency for Qatari authorities to opt for a gradualistic approach to problems is perhaps the most striking quality permeating Qatar's economic policies. This has enabled the country to avoid rushing into massive and unrealistic spending done simply in the name of development, and consequently largely to prevent such problems as inflation and congestion which have accompanied the oil boom in most petroleum countries of the Gulf. Institutionalisation of the Qatari financial market may be further explained thus: since the QNB had more experience as the agent of the government during the currency board era, the division of central banking functions between the QMA and the QNB is perhaps appropriate to the extent that both financial institutions are at least half-owned by the government. In any case time and experience are needed for the central monetary agency to acquire most of the characteristics of a full-fledged central bank capable of performing all the central banking duties.

The Present Money and Banking System

Some Structural Features

As noted, the most noticeable characteristic of the financial system in Qatar is the absence of full control by a central authority. It has been pointed out that the Qatar Monetary Agency shares with a quasi-private commercial bank, Qatar National Bank, the function of agent for the government in financial matters, especially in dealings with foreign contractors. Also it must be observed that the money market of Qatar is still at a relatively undeveloped stage. In a broad way, the financial structure in Qatar may be represented visually by Figure 6.1.

Figure 6.1: Qatar: Financial Structure

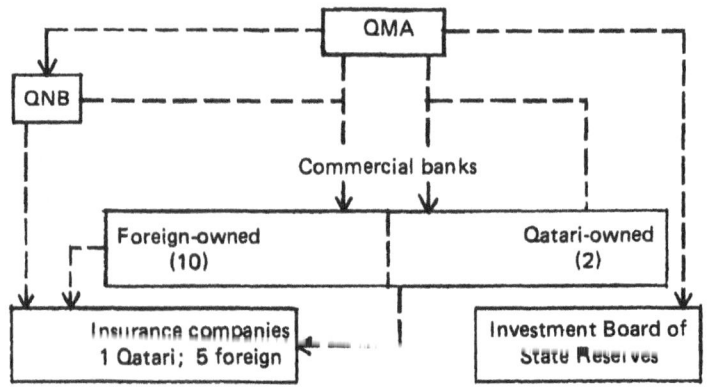

Note: Direction of arrows give rough indication of the flow of influence.

Yet another feature of the financial system is that the majority of the financial institutions are foreign-owned. Out of twelve commercial banks, only two are Qatari-owned, and five out of the six insurance companies in the country are owned by foreign interests. This may be a contributing factor to the sluggishness of any central monetary authority in exercising control over the financial system. It seems puzzling to observe that the foreign dominance of the financial system has come about despite the government's policy to restrict the operation of foreign banks. The reasons are that the restrictions came after most of the foreign banks had already been opened and that the objective of the restrictions is to ensure there is only one bank from each foreign country. Also, in a move to control the expansion of existing foreign banks, the 1970 Banking Decree prohibits the increase of branches in Qatar by foreign-owned banks.

The extent of capital flight out of Qatar in response to differential rates of interest is partially controlled by the requirements that foreign banks must have at least 50 per cent of their deposits in Qatar and, in addition, keep all their paid-up capital in the country. Moreover, expatriation of profits is restricted by the requirement of the 1970 law that 20 per cent of the annual profit of foreign-owned banks be retained in Qatar until authorised capital is reached.[9] However, these restrictions have not prevented private capital flight out of Qatar. Indeed, it may even be said that, reinforced by the statutory low-interest rates in the country, these restrictions might have encouraged capital flight. The non-banking private sector could borrow money at a low-interest rate (as low as 7 per cent) and re-deposit it abroad to earn interest as high as 14 per cent.[10] Also, Qatari-owned commercial banks are under fewer restrictions with respect to the manner in which they extend credit. It is not surprising, therefore, that a liquidity crunch in the Qatari financial market emerged as the pace of economic development quickened.

The Working of the System

Notwithstanding the still developing nature of the financial system, its foreign domination, and the liquidity shortage, the money and banking market of Qatar has expanded at an accelerated rate. Table 6.1 offers some statistics on the activities of commercial banks in Qatar over the period 1966-77. To the extent that the commercial banks constitute the bulk of the financial market in Qatar, the growth of their operations may be used as a measure of development in the nation's financial market. From a modest amount of QR 204.4 million in 1966, total private deposits held by commercial banks in Qatar increased to QR3,160.7 million at the end of 1977 — a more than fifteenfold expansion in deposits. It can be noted from Table 6.1 that credit expansion in the Qatar money market is even greater. Between 1966 and 1977, commercial banks' claims on the private sector in Qatar jumped from QR 103 million to QR 2,463.7 million, or approximately a 24-fold increase! An interesting observation is that the rate of expansion of credits, deposits, and other forms of financial instruments was greater in the second half of the period under discussion. For whereas both total private deposits and credits in the private sector by the commercial banks just about tripled from 1966 to 1972, by the end of 1977 the value of credits and deposits was respectively more than seven and five times the corresponding figures at the end of 1972.

The rapid expansion of the domestic activity of commercial banks in Qatar did not occur at the expense of their lending operations abroad. Given the low local interest rate which encouraged borrowing at home and re-depositing abroad, this should not be unexpected. Thus, despite the restrictions mentioned earlier placed on the overseas operations of the foreign commercial banks, foreign assets of the banks increased more than eightfold between 1966 and 1977.

To give an idea of recent developments, one may compare the commercial banks' statistics for the first quarter of 1976 and 1977 and February 1978. Table 6.2 offers the necessary information. The annual growth rates in all the financial variables in Table 6.2 (see the last column for each year) are very impressive for 1977 and considerably lower for February 1978. The relatively low rate at which net foreign assets grew between 1976 and 1977 (28 per cent) compared with 80 per cent gross foreign assets is an indication of an even more rapid rate at which commercial banks in Qatar incurred foreign liabilities. Perhaps this was in response to the liquidity crunch the Qatar money market has recently experienced. The February 1978 figures give an indication of a slowdown, although it is likely March 1978 figures may show a different picture.

Returning to Table 6.1, it appears that the reserve-total private deposit ratio kept by commercial banks has been very low, as low as 1.8 per cent at one time. The commercial

Table 6.1: Commercial Banks' Statistics, 1966-77 (end of year, in QR million)

	1966	1967	1968	1969	1970	1971	1972	1973	1974	1975	1976	1977
Reserves	9.5	5.7	7.8	9.7	6.6	9.5	16.2	17.8	29.9	57.7	42.0	77.2
Foreign assets	210.2	140.9	153.0	188.3	213.9	280.7	319.8	325.6	564.8	1,108.7	1,511.3	1,808.6
Foreign liabilities	14.9	17.4	9.2	15.8	39.7	25.7	2.5	31.2	103.6	225.1	345.9	478.2
Claims on private sector	103.0	113.1	154.1	182.9	214.7	262.4	332.7	503.4	752.2	1,125.6	1,559.1	2,463.7
Demand deposits	92.5	106.4	123.1	128.3	143.3	176.8	268.8	302.1	406.0	765.0	1,199.0	1,582.1
Time and savings deposits	111.9	125.0	123.4	188.0	187.0	225.8	318.6	364.2	485.1	743.7	1,128.9	1,578.6
Government deposits	106.3	22.2	23.4	9.5	11.7	55.8	20.2	39.0	126.7	229.2	119.8	153.3
Total private deposits	204.4	231.4	246.5	316.3	330.3	402.6	585.4	666.3	891.1	1,508.7	2,328.8	3,160.7
Capital accounts	19.5	17.9	25.5	31.6	53.2	66.6	67.2	95.2	168.9	252.0	210.1	232.3
Reserve-total private deposit ratio (%)	4.6	2.5	3.2	3.1	2.0	2.4	2.8	2.7	3.4	3.8	1.8	2.4
Reserve-demand deposit ratio (%)	10.3	5.4	6.3	7.6	4.6	5.4	6.0	5.9	7.4	7.5	3.5	5.0

Source: IMF, *International Financial Statistics*, vol. xxxi, no. 5, May 1978, pp. 324-5.

Table 6.2: Commercial Banks' Statistics for the First Quarters of
1976, 1977 and February 1978 (in QR million)

	I, 1976	I, 1977		Feb. 1978	
		Amount	Change %	Amount	Change %
Claims on private sector	1,206.3	1,827.6	52	2,349.8	29
Demand deposits	917.8	1,407.7	53	1,592.1	13
Time/saving deposits	713.3	1,281.6	80	1,553.7	21
Total private deposits	1,621.1	2,689.3	66	3,145.8	17
Foreign assets (gross)	1,202.9	2,170.3	80	1,797.2	−17
Net foreign assets	996.4	1,275.9	28	1,340.6	5

Source: IMF, *International Financial Statistics*, vol. xxx, no. 9, September 1977, p. 301; vol. xxxi, no. 5, May 1978, p. 325. Computation by ICEED staff.

banks in Qatar have been able to keep these ratios, despite the nonexistence of a lender of last resort, because of the access which they have to the financial markets in their metropolitan countries. Therefore, since it is easy to call upon these foreign financial markets for funds at short notice, it has not been found economical, given the liquidity shortage, to keep large internal reserves.

Activities of Qatari-Owned Financial Institutions

Despite the fact that, until April 1975, only one of the eleven banks operating was owned by Qatar interests, the share of the banking business in the hands of foreign firms has not been all that substantial. The reasons are basically two. First, the Qatar National Bank, due to the large government financial transactions which it handles, accounts for about 40 per cent of all banking business. Second, since the establishment of the Commercial Bank of Qatar (CBQ) — the first commercial bank under complete ownership of private Qatari interests — its activities have grown astronomically. The balance sheet of the CBQ on 30 September 1976, which stood at QR 122.7 million, showed a doubling over the previous nine months. Describing the factors which might have contributed to the phenomenal growth of this commercial bank, a special report of the *Middle East Economic*

Digest has said:

> . . . it is the only commercial bank in Doha offering a full
> foreign exchange service. But CBQ also has the advantage
> of being the only locally incorporated bank fully in private
> hands. Its comparative aggressive marketing style has
> helped to capitalise on its local connections and pull in
> both depositors and borrowers.[11]

The most important element is the ability of the local
bank to take advantage of the tendency of a segment of the
population who prefer overdrafts to loans, along with the
acceptance of personal security rather than other collaterals
akin to western developed financial markets.[12]

As will be recalled, the Qatar National Bank is considered
the pivot of the financial system, despite the existence of the
Qatar Monetary Agency. The growth of the QNB has been
subsidised by the government by exemptions from the 45 per
cent profits tax paid by all banks from 1971 until March
1976. Although the QNB is essentially the most visible financial agent of the government, the extent of its role in handling official foreign investment is limited. The Qatar General
Petroleum Company undertakes sizeable foreign investment
and the Qatar Monetary Agency holds foreign reserves to
back the riyal.

The overall investment strategy of Qatar rests in the hands
of the Qatar Investment Board, established by the Emir in
1972. The Board is headed by the Minister of Finance and
other members include the adviser to the ruler, the Director
of the Emir's private office, the Director of Finance, and a
Swiss banker. Recognising the complex nature of the foreign
investment market, the Board is believed to have a group of
advisers made up of international bankers among whom are
people from Manufacturers Hanover, First National Bank of
Chicago, Morgan Grenfell, and the Deutsche Bank.

The insurance business is dominated by the only indigenous company, the Qatar Insurance Company (QIC),
incorporated in 1964. The government's share of its QR 3
million capital is only 20 per cent. With the increase in
general economic activity, especially trade and infrastructural
activity, there has been a parallel boom in business for this

local insurance company. Its premium income almost tripled from QR 9.6 million in 1973 to QR 25.4 million in 1975. It is estimated to have reached QR 50 million in 1976.[13]

Money Supply Demand, and Monetary Policy in Qatar

Monetary Policy in a Less-Structured Financial Market

In a financial market characterised by loose control by a central monetary authority, an unstructured market for financial instruments, openness of the economy, and dominance by foreign-owned banks, the traditional weapons of monetary control which are available in more developed and well-organised money markets cannot be expected to apply. But, although legal reserve-deposit ratio, discount rate, and open-market operations have not been used consciously in Qatar as instruments of monetary or credit management, it will be farfetched to conclude that the Qatar government has no way of undertaking expansionary or deflationary policies if and when it chooses to do so. The government could simply increase or decrease spending from oil surpluses to achieve desired policies regarding the level of economic activities. The policy of the Qatar government to date, on the other hand, has been based on a gradual approach towards economic development in order to control inflation. 'It is clear that the government is determined to avoid the boom-bust cycle of hyperinflation followed by overcapacity and recessionary cooling which has occurred in other Gulf states.'[14]

From the nature of the policy described above it is obvious that the government of Qatar may undertake a monetary policy only if it is prepared to implement a corresponding fiscal policy. This can be done by the government causing the reduction (or expansion) of the major source of money supply, i.e., government spending.[15] Meanwhile, it may be observed that this is synonymous with increasing (or reducing) government deposits. The sources of money supply in Qatar will be discussed later. It is of interest now to turn to a brief monetary analysis, with a view to identifying what policy variables could be controlled by the government in order to manage money and credit in the economy.

Monetary Analysis

Table 6.3 gives a convenient method for carrying out monetary analysis in Qatar. Money supply, defined in the narrow sense to include currency and demand deposits (or designation M1 in the table), rose from QR 117.2 million in 1966 to QR 2,084.2 million in 1977, an increase of over 1,678 per cent. On the other hand, according to the broader definition of money (M2), money supply in Qatar increased by about 1,499 per cent, 179 per cent points less than M1. This means time and savings deposits grew more slowly than the combined growth of other components of money. Consequently, although money and quasi-money combined had a substantial growth between 1966 and 1977, it appears that the portfolio composition of the Qatari private sector shifted from time and savings deposits to more liquid forms of assets. In particular, there was a tendency for demand deposits to be preferred to the other liquid assets, as evidenced by the increasing proportion of total money supply which demand deposits constitute. Currency in the hands of the general public recorded the slowest growth rate among the three components of money, followed by time and savings deposits.

As often happens in an expanding economy which concurrently experiences an increasing level of monetisation and rapid economic development, the currency-deposit ratio in Qatar fell from 43.7 per cent in 1967 to 31.1 per cent in 1976 and 31.7 per cent in 1977. Consequently, the oft-quoted characteristic of Qataris being 'extremely cash conscious',[16] though it might have been true in the 1950s, is certainly no longer valid. In fact, when one compares the Qatar statistics with 211.6 and 78.8 per cent in 1970 and 1976, respectively, for Saudi Arabia, it is clear that Qatar is not as cash oriented as has been assumed generally to be the case. A further examination of the comparative currency-demand deposit ratios of Qatar and the United States at the end of 1976 (31.1 per cent for Qatar and 34.3 per cent for the United States) indicates, quite unexpectedly, that Americans are more cash conscious than Qataris.[17] This is a rather surprising revelation. However, considering the fact that the business community constitutes the dominant sector in terms of monetary holdings in Qatar, the low currency-deposit ratio should not be unexpected given the business sector's

Table 6.3: Qatar: Monetary Survey, 1966-77 (end of year, in QR million)

	1966	1967	1968	1969	1970	1971	1972	1973	1974	1975	1976	1977
Currency	24.7	46.5	50.9	51.5	54.3	60.3	76.7	110.7	154.6	236.9	373.2	502.1
Demand deposits	92.5	106.4	123.1	128.3	143.3	176.8	268.8	302.1	406.0	765.0	1,199.9	1,582.1
Money supply (M1)	117.2	153.0	173.9	179.8	197.6	237.1	345.5	412.8	560.5	1,001.9	1,573.1	2,084.2
Quasi-money	111.9	125.0	123.4	188.0	187.0	225.8	318.6	364.2	485.1	743.7	1,128.9	1,578.6
Money supply (M2)	229.1	278.0	297.3	367.8	384.6	462.9	664.1	777.0	1,045.6	1,745.6	2,702.0	3,662.8
Increasing factors	332.6	288.9	363.3	431.0	473.6	611.8	775.8	1,095.7	1,496.1	2,425.6	3,266.9	4,464.4
Foreign assets (net)	229.6	175.8	209.2	248.1	258.9	349.4	443.1	592.3	743.6	1,300.0	1,707.8	2,000.7
Claims on private sector	103.0	113.1	154.1	182.9	214.7	262.4	332.7	503.4	752.5	1,125.6	1,559.1	2,463.7
Decreasing factors	103.5	11.0	66.0	63.1	89.1	148.9	111.8	318.8	450.5	679.9	565.0	801.6
Government deposits	107.3	22.2	23.4	9.5	22.7	64.7	44.0	68.9	171.1	338.4	266.3	253.7
Other factors (net)	−3.8	−11.2	42.6	53.6	66.4	84.2	67.8	249.9	279.4	341.5	298.7	547.9

Source: As for Table 6.1.

practice of cheque transactions.

Money supply, in whatever form, has certainly grown at an accelerated rate in Qatar during the 1970s. Factors which have contributed to this rapid growth, as evidenced in Table 6.3, are (a) the accumulation of foreign assets, and (b) more importantly, increases in loans and advances given to the private sector. The banking system in Qatar has therefore responded to the credit needs of the expanding economy by making credit available more readily to private businessmen and households despite the liquidity shortages which have recently beset the financial market.

The liquidity shortage may be explained by the growth of government net deposits with the banking system. The rapid increase in this item in the monetary analysis (Table 6.3) had a retarding effect on the growth of money supply, including quasi-money. This is seen in the fact that whereas between 1969 and 1976 M2 grew by 635 per cent, government net deposits recorded a rate of growth of 752.4 per cent, while other decreasing factors grew by 588.3 per cent (Table 6.3).

Conclusion

The analysis in the foregoing pages has traced the long history of money and banking in Qatar, pre-dating independence. From a system of foreign currency circulation to the present arrangement of a loose form of central monetary agency, the economy has become increasingly monetised. Monetary development in Qatar by and large has been accelerated by the oil boom in the 1970s, which has enabled money supply to grow at a fast rate despite the deflationary impact of the government's accumulated surpluses.

The rapid growth of government net deposits with the banking system means that governmental spending has been held below the level of accruing revenue. The government is capable therefore of controlling the growth of the money supply by managing its expenditures, i.e., accumulating deposits. Thus the tools of monetary/fiscal policy are available to the authorities. The analysis has also underscored the contention that monetary and fiscal policies in Qatar are very much interwoven.

As the Qatar Monetary Agency acquires more experience, other instruments of monetary control will become available and will be utilised to make monetary policies more independent of fiscal measures. Meanwhile, it may be concluded that the authorities in Qatar have adopted moderate and cautious monetary/fiscal policies with the result that inflation has been kept to the minimum.

Notes

1. See Michael E. Edo, *Currency Arrangements and Banking Legislation in the Arabian Peninsula* (International Monetary Fund, Middle Eastern Department DM/74/86, 30 August 1974).

2. Ibid., p. 7.

3. Until recently, the only indigenous bank in Qatar was the Qatar National Bank, established in June 1964.

4. Michael E. Edo, *Currency Arrangements*, pp. 8-9.

5. For example, see E.K. Hawkins, 'The Growth of Monetary Economy in Nigeria and Ghana', *Oxford Economic Papers*, vol. 10, no. 3, October 1958.

6. It may be noted that, even after the establishment of Qatar Monetary Agency, many of the central banking functions are performed by the Qatar National Bank, which is half-owned by the government.

7. Michael Edo, *Currency Arrangements*, p. 6.

8. See appendix for excerpts from the Banking Law of 1973.

9. N.A. Shilling, *Doing Business in Saudi Arabia and the Arab Gulf States* (Inter-Crescent Publishing and Information Corporation, New York, 1975), p. 268.

10. Ibid., p. 269.

11. MEED, special report, April 1972, p. 12.

12. Ibid.

13. Ibid., p. 13.

14. *Mideast Markets*, 10 April 1978, p. 2.

15. Details of the government fiscal and other related policies have been discussed in the chapter on public finance.

16. MEED, special report, April 1977, p. 12.

17. All the currency-deposits ratios used in the discussion in the paragraph were computed from a single source, IMF, *International Financial Statistics*, vol. xxxi, no. 5, May 1978.

Appendix

The Qatar Monetary Agency

Legal Basis. Monetary Agency Law, 1973 (19 May 1973).

Name. There shall be established an Agency to be known as the Qatar Monetary Agency. Article 3 (1).

Purposes of the QMA. The purposes of the Agency shall be: (a) to regulate the issue, redemption and supply of currency; (b) to safeguard the internal and external value of the currency and its convertibility into other currencies; (c) to promote credit and exchange conditions conducive to the growth of the economy within the framework of monetary stability; and (d) to supervise the banks. Article 4.

Board of Directors. (Summary of provisions) The board consists of a Governor, a Deputy Governor and five other Directors. Members are appointed by the Emir on the recommendation of the Minister responsible for Finance Affairs, the Governor and Deputy Governor for five-year terms, and the five other Directors for three-year terms. Appointments are renewable. The Governor and Deputy Governor must be full-time officers; the Governor is the Chief Executive and the Deputy Governor deputises for him in his absence.

Capital and Ownership. The capital of the Agency shall be ten million riyals and shall be paid by the Government in full upon the establishment of the Agency. Article 6 (1).

Currency and External Reserve Requirements. The unit of currency of Qatar shall be the riyal and shall be divided into one hundred dirhams. Article 21.

The par value of the riyal and any change thereof shall be declared by an Emiri decree on the recommendation of the Agency and the approval of the Minister. Article 22.

The value of the reserve of external assets provided for in paragraph (1) of this Article shall not be less than an amount equivalent to 100 per cent of the currency in circulation provided that the Board may, from time to time, with the approval of the Minister, alter the minimum reserve figures for periods not exceeding six months. Article 29 (2).

See page 182 Article 29 (1) specifies the permissible external reserve assets.

Definition of 'Bank' and 'Banking Business'. 'Bank' means any company licensed to carry on banking business in Qatar under Part IX of this law. Article 2 (c).

'Banking business' means the business of accepting from the public deposits repayable on demand or otherwise and withdrawable by checks, draft, order or other means, for the purposes of extending credit or investment at the risk of the person accepting such deposit. Article 2 (d).

Head Office, Branches, Agents and Correspondents. The Agency shall have its Head Office in Doha and may, within Qatar, or abroad: (a) establish and close branches; (b) appoint agents and correspondents and cancel such appointments. Article 5.

7 INTERNATIONAL LINKS: TRADE, AID, INVESTMENT AND REGIONAL COOPERATION

Introduction

Although Qatar is small in area and population, it has considerable international links. In the past, these flowed from the country's location, maritime orientation, and traditional economic activities. Today, the long-established patterns have been extended and widened due to Qatar's dual and inter-related status as an oil exporter and a capital-surplus nation. These regional and global connections are forged primarily in the economic areas of trade, investment, and assistance programmes.

International trade, based overwhelmingly upon the export of oil, is the mainstay of the Qatari economy. The petroleum sector alone provides the major contribution to gross domestic product (GDP) and is the primary source of government revenues and foreign exchange.

In spite of the capital surpluses derived from the oil sector, Qatar still faces many of the real challenges of development. International trade theoretically enables a country to export those commodities in which it has a comparative advantage in production, in exchange for goods and services produced abroad at lower costs. Given certain constraints in the case of Qatar, the greater the returns from trade, the greater are the benefits of economic growth and progress towards economic development.

In addition to the static gains from trade, e.g., foreign exchange savings, transactions within the foreign sector will provide further benefits, such as capital goods, technical and managerial skills and services, and other infrastructure requirements the fulfilment of which are indispensable to the country's economic development. One should keep in mind the particular gains from trade and the effect of these gains on the Qatari economy, as well as the overall significance of trade for regional cooperation and development.

One of the objectives of Qatar's development policy is to coordinate the national industrial programme with the overall

139

development of the Gulf region. The potential gains from trade and regional cooperation between Qatar and the rest of the Arab Gulf states transcend the purely economic realm and may include almost every aspect of national life. There is a strong tendency in the region toward some sort of political coordination or degree of integration, most visible in the establishment of the United Arab Emirates.

The Structure of Qatar's International Trade

The Development and Composition of the Imports

Imports to Qatar are closely associated with levels of oil revenue, population growth, and the state of economic activity, in particular the level of development expenditure. Due to the limited agricultural and industrial base of the economy, the country is highly dependent on imports to satisfy rising local demand for consumer goods as well as intermediate and capital goods.

Table 7.1 illustrates the expansion of Qatar's imports since 1960. Despite the fact that the nation's imports have increased more than tenfold since 1970, import expenditure accounted for less than one-third of the total government revenue derived from oil in 1976. Present and future trends have been accentuated by two important factors affecting government revenue: (a) the 1973-4 oil price increases and subsequent price rises; (b) the government's recent assumption of 100 per cent participation in foreign oil company operations in Qatar.

While Qatar's industrial development programmes have been major items of expenditure over the past few years, imports of consumer goods remain substantial. For example, consumer goods and durables accounted for just under 60 per cent of the value of total imports in 1969; in two years, the share of these classifications declined to about 45 per cent (1971).[1]

The import picture denotes: first (Table 7.1), the value of imports to Qatar began to climb beginning in 1967. This is attributed to a considerable extent to the steady rise in oil production and revenues since 1967. The sudden jump of imports in 1971 and 1974 can be explained by the sharp increases in oil revenues in these two years.

Table 7.1: Imports to Qatar, 1960-76 (c.i.f. in millions QR)

Year	Value of Imports	Year	Value of Imports
1960	134.7	1969	252.4
1961	110.9	1970	305.5
1962	139.9	1971	516.1
1963	116.9	1972	607.4
1964	119.1	1973	778.3
1965	124.6	1974	1,068.9
1966	124.4	1975	1,621.6
1967	138.3	1976	3,290.4
1968	201.1		

Sources: State of Qatar, Ministry of Economics and Commerce, *An Analysis of the Commodity Imports to Qatar* (Dar El-Ulum Establishment, Doha, 1972), pp. 18-19; International Monetary Fund (IMF), *International Financial Statistics* (IMF, Washington, DC, September 1977), pp. 300-1.

Second, all import categories have registered gains although the various categories reflected different rates of growth. The drop in the share of consumer goods and durables was noted earlier; in the same 1969-71 span, imports of capital goods quadrupled.

This steady rise in imports is the result of the interaction of many forces: the availability of foreign exchange through the high and continuously rising oil revenues, the limited agricultural and industrial base of the Qatari economy and thus the high average and marginal propensity to import, the rapid increase in population, and most importantly, the vigorous efforts at economic diversification, social development, and industrialisation. The development and industrialisation policies, moreover, have contributed to the shift in the composition of imports which has moved from consumer goods towards capital goods.

Table 7.2 presents the geographical distribution of Qatari imports over the four-year period 1972-5. As suggested by the table, Western Europe traditionally has held the lion's share of trade with the state of Qatar, followed by Asia and the Middle East. The pace of economic development has caused trade to expand beyond the Middle East to regions

Table 7.2: Geographical Distribution of Import Trade to Qatar, 1972-5

Region	1972		1973		1974		1975	
	QR million	Per cent	QR million	Per cent	QR million	Per cent	QR million	Per cent
Western Europe	290.4	48.2	370.8	46.7	372.6	35.5	758.1	47.1
of which:								
United Kingdom	160.6	26.7	214.9	27.1	149.7	14.3	342.3	21.3
West Germany	32.8	5.3	42.2	5.3	65.8	6.3	150.7	9.4
Eastern Europe	3.2	0.5	6.5	0.8	13.3	1.3	15.6	1.0
North America	63.4	10.5	80.5	10.1	123.8	11.8	205.1	12.7
of which:								
USA	63.2	10.5	80.0	10.1	109.5	10.4	201.6	12.5
Brazil	—	—	16.3	2.1	—	—	—	—
Middle East	109.5	18.2	142.5	17.9	192.5	18.4	222.8	13.8
of which:								
Lebanon	42.0	7.0	44.2	5.6	67.0	6.4	64.7	4.0
Africa	4.8	0.8	3.9	0.5	9.2	0.9	13.2	0.8
Asia	112.9	18.7	137.7	17.3	303.9	29.0	358.7	22.3
of which:								
Japan	72.1	12.0	86.4	10.9	190.9	18.2	242.2	15.1
Oceania	18.4	3.1	23.2	2.9	33.0	3.1	34.5	2.1
Other	—	—	13.0	1.7	0.3	—	1.9	0.2
TOTAL	602.6	100.0	794.4	100.0	1,048.6	100.0	1,609.9	100.0

Source: Customs Department, *Yearly Bulletin of Imports, Exports and Transit* (State of Qatar, 1977).

Table 7.3: Commodity Composition of Qatar Imports from United Kingdom and United States (in percentages)

Commodity	UK 1975	UK 1976	US 1976
Food and live animals	1.6	1.9	1.8
Beverages and tobacco	1.6	1.8	0.5
Crude materials (inedible)	0.2	0.1	0.6
Minerals, fuels, lubricants, etc.	1.2	1.7	0.3
Animal and vegetable oils and fats	0.1	0.1	0.1
Chemicals	5.5	4.9	3.7
Manufactured goods	24.3	20.7	7.5
Machinery and transport equipment	57.8	60.2	79.9
Miscellaneous manufactured goods	7.7	8.6	5.6
Total	100.0	100.0	100.0

Source: MEED, special report, April 1977, p. 31.

where intermediate and capital goods are more readily available. Moreover, five countries distributed over four major regions have accounted for approximately 60 per cent of all imports to Qatar; the United Kingdom remains Qatar's major supplier, followed by Japan, the United States, Lebanon, and West Germany.

A more recent indication of the commodity breakdown of Qatar imports is revealed by examining the export figures of the nation's major supplier: the United Kingdom (see Table 7.3). The trade emphasis in 1975 and 1976 clearly has been on machinery and transport equipment, followed by manufactured goods. This trend is further reinforced by the fact that 80 per cent of United States exports to Qatar in 1976 consisted of machinery and transport equipment.[2] The stress on capital goods imports to Qatar will continue to mirror the level of governmental spending on development programmes.

With respect to the export side of the trade balance, the dominant feature is that essentially all crude oil produced in Qatar is exported. In 1971, Western Europe as a region received nearly 70 per cent of Qatar's exported petroleum, with the United Kingdom, West Germany, and France purchasing 17.8, 16.9, and 13.8 per cent, respectively (as traced in the earlier chapter on the oil industry). By 1975, the direction of export trade had altered slightly, mainly towards the United States. In that year the United States purchased more than 20 per cent of Qatar's crude oil production, thus becoming the largest single importer of Qatari oil. While Western Europe remained the major regional importer (taking 55 per cent of Qatar's crude export market), the Netherlands ranked second among the individual importers. The balance of trade with Qatar's second major trading partner — Japan — has continued to tip in favour of that country. Throughout the 1970s, Japan has held less than one per cent of Qatar's petroleum export market.

In the decade prior to 1973, Qatar enjoyed a favourable balance of trade in all but two years (1970 and 1971). Although this trend is expected to continue in the future, it should be recalled that the balance of trade does not reflect the total or accurate financial or liquidity position of the country, since it neither includes capital transactions nor other invisibles. Capital transactions are especially critical to the Qatari economy, in part because Qatar maintains considerable investment abroad and, moreover, because the majority of the expatriate population remits a significant portion of its earnings to its home country. A brief discussion of Qatar's balance-of-payments position would be useful here.

Balance of Payments

Table 7.4 sketches a brief view of Qatar's balance-of-payments position for the period 1973-6. Various factors should be noted. First, payments received from crude oil exports nearly quadrupled from 1973 to 1974, while the actual volume of petroleum exported decreased by nearly 10 per cent.[3] This, of course, is the direct result of the OPEC price increases in this period. The decrease in total export revenues in 1975 coincided with further decreased production due to stockpiling, the beginning of a temporary glut in the world market, global economic recession, and a mild winter in the West, all of which contributed to decreased demand for Qatar's crude.

Second, the proportion of non-oil exports to oil exports has been changing. Since 1974, other exports have assumed a faster growth rate than oil exports. The implications of this

Table 7.4: Qatar's Balance of Payments (in millions of US dollars)

	1973	1974	1975	1976[a]
Oil exports, f.o.b.	598	1,955	1,763	2,131
Other exports, f.o.b.	18	37	51	72
Imports, c.i.f.	−195	−271	−458	−817
Services and private transfers	−140	− 53	−227	−348
Current account balance	280	1,667	1,130	1,039
Capital and official transfers[b]	−283	−613	−527	−720
Overall balance	− 3	1,054	603	318

[a] Preliminary figures.
[b] Includes errors and omissions. Capital and official transfers are on a fiscal year basis; other items are on a calendar basis.

Source: IMF, *Survey*, 15 August 1977, p. 259.

are important for the prospects of sustainable economic growth in the future. Moreover, they are an indicator of the relative impact of the government's objective of economic diversification.

Third, the quadrupling of payments for imports over the three-year period (1973-6) is primarily the result of economic development and the need for goods and services of all kinds. The major share of this increase may be attributed to the demand for capital goods imports required for industrialisation as discussed previously.

Qatar and the Potential Regional Framework

Qatar's interest in the region — both the Gulf and the larger arena of the Arab Middle East and North Africa — is deep-rooted. The entrepôt characteristic commonly found among such Gulf states as Kuwait, the United Arab Emirates, and Bahrain, also contributed to Qatar's outward-looking attitude and trade-oriented activities. The country's resource constraints (other than petroleum) and cultural ties combined to strengthen the economic basis of regionalism in the past. With the advent of the oil period, the accelerated development programmes and petroleum logistics have widened the regional perspective still further.

Qatari requirements in labour are largely met from the

area; because its size dictates a narrow domestic market, Qatar needs markets for its growing and projected spectrum of industrial products and again, the region may absorb some of this output. The generator of Qatari government revenue, that is, the export of crude oil, is closely linked with petroleum transport through such means as Egypt's Suez Canal and Sumed (Suez to Mediterranean) pipeline. As Qatar moves into refining and other downstream oil industry operations, including petrochemicals, the Canal assumes even more importance since product carriers are considerably smaller than the crude tankers and thus can easily use that waterway.

To labour needs, markets, petroleum logistics, and cultural ties should be added Qatar's very real concern for the economic and political stability of the region as well as its orderly development. The degree and form of regional cooperation and/or coordination which Qatar may seek or eventually employ reflect a healthy range of options. A number of measures, often on an ad hoc basis, have been instituted in the Arab world and within the Gulf region such as the Arab Monetary Fund, the Organization of the Arab Petroleum Exporting Countries (OAPEC), regional multinational aid agencies, and the proposed Gulf Common Market; perhaps these undertakings can be better evaluated if some of the theoretical aspects of regionalism are briefly reviewed.

There are many forms, phases, and degrees of economic integration. A free-trade area, for example, abolishes tariffs and other trade restrictions among members, while at the same time each nation is able to maintain its own tariffs on imports relating to non-members. In short, the bloc does not adopt a unified tariff policy on imports from outside. A customs union goes one step further by maintaining a common tariff on trade with non-members. In a common market, members abolish restrictions on both goods and factor movements among themselves besides applying a common tariff policy vis-à-vis the outside world. An economic union is a common market plus a programme designed to coordinate national economic policies. Total economic integration involves the establishment of a supranational agency with the power to decide the nature of a common monetary, fiscal, and socio-economic policy for the whole membership.

One of the principal reasons for forming any type of economic integration is to give preferential tariff treatment to goods imported from the bloc forming the union, and thus to discriminate against goods imported from outside the union. Hence, a discriminatory tariff policy, while disturbing existing economic relations, would cause significant trade diversion and some trade creation.

> If it shifts production from a higher-cost source to a lower-cost source, it creates trade and moves toward the free trade position. If the shift is from a lower- to a higher-cost source, it diverts trade and moves away from the free trade position. Such shifts in production constitute the allocation effect of a customs union — they affect the static efficiency of resources use.[4]

The welfare implications of a discriminatory tariff policy are not confined to production, but also include some consumption effects resulting from the probable increase in imports. As prices decline when trade is diverted from non-member to member countries, the consumption of the respective goods increases. The rise of imports in excess of the amount of diverted trade may be labelled trade expansion.[5]

In addition, the net effect on living standards depends essentially on the magnitude of trade creation relative to trade diversion. The higher the trade creation relative to trade diversion, the greater the potential gain from integration will be. The conditions that tend to augment trade creation and limit trade diversion may include: (a) high initial structure of tariffs for union members relative to tariffs of outside nations, (b) high demand elasticities by union members for each others' goods, and (c) low elasticities of demand by union members for outside goods.

The desirability of economic integration also depends on its effect on welfare. In a situation which involves many nations, interest might be focused on the welfare of one nation, member or non-member, or alternatively of the union vis-à-vis the outside world. Strictly speaking, the interest of the union tends to be in conflict with that of outside countries. Thus, rigid conformity with the principles of modern welfare theory forestalls any conclusion as to the net welfare effects of regional blocs.

One of the basic conditions that tends to contribute to greater trade creation is the extent of trading between union members before the formation of the economic bloc.[6]

> The greater the proportion of preunion trade of each member-to-be, that is, with other members-to-be (i.e., the lower the proportion of preunion trade with countries not included in the union) the greater is the probability that the union will raise welfare.[7]

Furthermore, the larger the economic size of the union, the greater will be the ensuing benefits of the reallocation of production and the wider the scope for specialisation due to an increased market size and the availability and range of resources. In addition, as the size of the union increases, the incidence of trade diversion is reduced.

It has been stated generally that an economic bloc among complementary economies is apt to realise substantial net beneficial effects in contrast to one among rival economies. The argument is based on the notion that complementary economies possess different patterns of resources, and thus production, so that integration may foster specialisation in those goods each country is best suited to produce.

This view has been contradicted by maintaining that integration between rival economies is more likely to be beneficial because, in this case, trade diversion would be at a minimum.[8] This may be true, but since rival economies tend to have similar cost ratios, integration among them cannot result in significant savings despite the fact that trade creation does occur.[9]

Theorising along these lines, however, does not seem at this juncture to be very relevant to most developing countries such as Qatar, since the primary objective of forming an economic bloc in such a setting is the creation of potentially efficient industries and not the increase in the efficiency of existing industries (as in advanced economies). Viewed in this light, the problem becomes part of the dynamic aspect of the formation of regional blocs.

Dynamic gains of integration result from the exploitation of large-scale economies, which comprise increased efficiency evolving from internal economies, from potential acceleration of technological progress, and possibly from lessened uncertainty.[10]

Since the amount of investment is restricted by the size of the market and because most modern industries either require a relatively large minimum plant size or are characterised by increasing returns to scale (at least up to a certain point), the formation of a regional bloc by enlarging the market would facilitate the establishment of large-scale, modern, and efficient industries. These industries would produce external economies in the form of high demand for different goods and services, on-the-job training for a portion of the labour force, and modernisation of the value system (among other elements), thereby furthering economic and social development.

The formation of a regional bloc may also change the structure of the market and thus either increase or decrease competition. If internal economies are important, integration tends to reduce the number of firms producing a specific commodity for the whole union market, which results in increased concentration. Other things being equal, a higher degree of concentration will diminish the effectiveness of competition.

Others think, on the contrary, that integration tends to undermine monopolies. They feel that since the removal of tariffs tends to increase the number of potential competitors it diminishes the market power of the domestic producers, although the union as a whole may have fewer firms because of economies of scale. It has been argued that by inducing the firm to grow, integration will cause rapid technological change as there is a direct relationship between the size of the firm and its outlays on research and development. Furthermore, the research and development industry is characterised by large-scale economies that could contribute to a higher rate of technological progress. The conclusion is that the exploitation of economies of scale permitted by integration will accelerate technological advances and thus lead to a higher rate of economic growth.[11]

The creation of a regional unit could contribute significantly to a greater stability of market and thereby eliminate, or at least reduce, a major risk in international trade that arises from changes in tariffs, quotas, subsidies, exchange rates or exchange restrictions.[12] This gain is especially critical for nascent industries in developing countries which aspire to further their industrialisation through integration.

Potential Economic Benefits of the Proposed Gulf Common Market

Almost all Arab states in the Gulf seem to recognise the fact that some degree of regional integration could serve their goals of social and economic development. This was recently expressed in a proposal for a Gulf Common Market.[13] It is of interest to analyse, in the light of theoretical discussion of integration, the potential gains or costs of such an arrangement from the participant's point of view.

It appears that the conditions which tend to make any kind of economic integration viable, with an eye to static resource allocation theory, are not prominent. The population base of the proposed union is very small. In addition, these countries do not engage in much trade among themselves and the demand elasticities for each others' products are low while their elasticities are high with respect to the products of outside countries. Furthermore, the initial structure of their tariff is low. In the event of a common market being established, these conditions will cause substantial trade diversion and thus diminish the welfare of the member countries.

However, it must be emphasised that, at present, these countries produce certain petrochemical and other energy-intensive products mainly destined for the export market. As noted earlier, the profitability of such industries depends significantly on the degree of valorisation of the respective products. A common market arrangement could contribute to the coordination of production and research policies, the intensifying of market surveys, the enhancement of bargaining power, and the lessening of harmful competition between member countries, thereby increasing the profitability of these industries. A common market arrangement is not strictly necessary for the adoption of such policies as similar results can be reached by active coordination on the sectoral level.

Nonetheless, the main gains from a common market arrangement would derive from its dynamic effects. It has been pointed out that efforts at diversification and import substitution are hindered, in part, by the narrowness of each country's domestic market — a situation that does not permit the exploitation of economies of scale. A Gulf Common Market would certainly widen the domestic market and encourage the establishment of more efficient, import-substituting industries. Moreover, specialisation could be fostered in the region and wasteful duplication and competition could be avoided or reduced to the benefit of all members.

Although these countries enjoy similar resource endowments, there is some variation in the availability of certain resources that could be exploited advantageously. This is especially true with respect to agricultural resources. Ras al-Khaimah and especially Oman possess significant agricultural potential that could be developed, provided investment funds are available and a stable market secured.

It must be re-emphasised that the size of the proposed Gulf Common Market, in absolute population terms, remains small. It would probably be insufficient for the production of many intermediate and capital goods. Nevertheless, taking into consideration the extremely high per capita incomes of member countries of the proposed common market, the market, in value terms, may not be as small as the population size might indicate. In any case, due to the limited resource base, many consumer goods would still have to be imported, at least during the early years of the common market.

In conclusion, a common market would undoubtedly expand the economic opportunities and push the production possibility frontier outward, thus increasing the absorptive capacity of the economies of member countries; yet such an arrangement should not be seen as a panacea for all the more restrictive economic ramifications of a small size.

In the past, Sheikh Khalifa of Qatar has been a major force behind such a common market. Furthermore, a currency arrangement between Qatar and Dubai and the joint economic commission established with Bahrain are indicative of Qatari efforts towards regional economic integration. It should also be mentioned that the formation of some type of larger economic unit in the Gulf and/or Middle East does contain some spin-off in the political realm in that the smaller nation-states can, in a sense and if so desired, speak with a stronger voice in the global community.

Nor are coordination and cooperation in the Gulf limited to purely economic concerns. That body of water is an age-old trading route, a source of employment and income through fishing, and a gate through which vital shipments of crude oil and petroleum products pass daily. In April 1978, Qatar joined with seven other Gulf nations to approve two treaties on cleaning the Gulf and protecting it from further pollution and damage.

Hammered out in 18 months of negotiations under the auspices of the United Nations Environment Programme (UNEP), the treaties have achieved something of a record in the short time required for their formulation and acceptance. The international agency warned the countries involved in 1976 that, while not so heavily polluted as the Mediterranean, their Gulf still had severe pollution problems from oil. The Mediterranean's condition can be traced primarily to industrial waste and sewage from the land and, while this is not yet true in the Gulf, the rapid industrialisation there could soon make it so. For example, industrial investment along the Arabian coastline is presently about $40 million per kilometre and some $20 million per kilometre on the Iranian side of the Gulf — significant indicators of the pace of industrialisation in the region. Moreover, approximately 60 per cent of the oil in international trade carried via tankers passes through the Gulf (about 1 billion tons annually); the possibility of massive spills and accidents clearly exists. With an average depth of only 100 feet or so and being salty and warm, the Gulf's capacity to break up and absorb industrial waste and sewage is severely limited.

It is encouraging that the drive towards industrialisation in the immediate region has not been given an unchecked 'green light', so to speak, as is so frequently the case in developing countries. Admittedly, the oil states of the Gulf have sufficient capital to allow them to avoid an either-or quandary — either industrialisation or expenditure on environment. Nonetheless, the eight signatories agreed to strict regulations to halt the discharge of waste by vessels and aircraft and to control the dumping of waste from coastline industries into the Gulf. The treaties will set up a marine environmental protection agency (most likely in Kuwait) plus a possible service of pollution-fighting ships and aircraft, perhaps based in Bahrain. The Gulf states created a $ 6.3 million trust fund to underwrite the initial implementation of these treaties.

Additional areas in which Qatar has moved to promote the concept of regional cooperation include its support of the Arab Monetary Fund (AMF). Qatar has underwritten $ 35 million of the $ 875 million or so total to this body for the purposes of alleviating members' balance-of-payments deficits and currency stabilisation.[14] Created in 1975, the Fund is restricted to the 21 members of the Arab League with the enunciated goals of fostering Arab economic integration and development. The AMF will guarantee loans to Arab countries to correct deficits resulting from individual or pan-Arab development projects and thereby encourage more joint Arab economic planning. The Fund expects to function as a catalyst in the creation of Arab financial instruments and a unified Arab currency as well as ultimately becoming the clearing house for all Arab central banks. The strategy includes the issue of bonds for sale to Arab governments, with the proceeds being re-lent. The AMF is still in its early days but it remains an ambitious undertaking.

Among other Qatari regional links, some of which have been noted earlier, are: the United Arab Shipping Company; membership in the inter-Arab consortium which signed the Sumed Pipeline Accord in 1974; a proposed Qatari-Jordanian Bank; OAPEC-based enterprises, such as the Arab Tankers Company and the Arab Petroleum Investment Company; and a host of Arab League affiliations in areas relevant to communications and standards.

Foreign Aid and Development Assistance

Another facet of regionalism in the Middle East has emerged with the establishment of a plethora of institutions designed to implement programmes of foreign assistance for the benefit of the non-oil Arab states and other developing nations.[15] Although Qatar has no indigenous aid institution per se, the Qatari government has been a generous contributor to regional entities such as the Arab Fund for Economic and Social Development (AFESD), the Arab-African Oil Assistance Fund, the OPEC Special Fund, the IMF Oil Facility, and other international aid agencies, to name but a few. Through these aid institutions and the bilateral assistance extended on a government-to-government basis, Qatar has emerged at the forefront of donors worldwide in respect of the proportion of its GNP expended on such aid. From 1973 to 1976 (Table 7.5), Qatari assistance has represented an annual average of 12 per cent of the nation's GNP — a staggering share when compared with the record of the industrialised Western bloc which has been the traditional source of development and foreign assistance since the close of World War II.

Many will recall that when the United Nations proclaimed the 'development decade', the goal for the advanced, aid-extending nations was set at 1 per cent of the respective GNPs; not one of the OECD (Organisation for Economic Cooperation and Development) members attained that objective in their foreign assistance programmes. Sweden and Holland reached the highest (0.7 per cent), while the United States' economic assistance in recent years has not exceeded more than one half of one per cent of its GNP.[16] In fact, the only aid extenders meeting the 1 per cent guideline are the major oil exporters; in addition to Qatar, which leads with 12 per cent of its GNP going to foreign aid from 1973 to 1976, the other nations in this category for the same period are the United Arab Emirates (10.2 per cent of GNP), Kuwait (5.3 per cent), Saudi Arabia (about 5 per cent), Libya and Iraq (1.8 per cent).

In absolute terms and as expected, Qatar's level of assistance falls below that of the oil 'giants', such as Saudi Arabia, the United Arab Emirates, and Kuwait, which in the 1973-6 span extended $5.65 billion, $2.87 billion, and $2.47 billion, respectively. Nonetheless, the Qatari aid level for the period of $792.9 million is almost equal to that of much larger Iraq ($822.4 million) and surpasses the contributions of Libya ($726 million).[17]

Several characteristics become clear when studying the level, forms, and recipients of Qatari foreign aid. The first is that in bilateral lending and assistance (Table 7.6), the

Table 7.5: Summary of Qatari Aid, 1973-6 (in $ millions)

	1973	1974	1975	1976	1973-6 Total
Commitments	93.1	227.7	369.1	138.3	828.2
Multilateral	0.5	30.3	31.7	38.7	101.2
Bilateral	92.6	197.4	337.4	99.6	727.0
Net disbursements	93.7	184.2	338.9	175.1	791.9
Multilateral	1.1	15.8	36.2	31.1	84.2
Bilateral	92.6	168.4	302.7	144.0	707.7
Disbursements as % of GNP	15.6	9.3	15.6	7.4	
Annual average 1973-6 disbursements as % of GNP					12

Source: Compiled from Organisation for Economic Cooperation and Development statistics.

procedure is relatively uninstitutionalised and the agreements are not widely publicised. Primarily the loans are linked to specific development projects in the recipient nations (project lending): the improvement of the Ugandan transport network and in Egypt the reopening and expansion of the Suez Canal and the Talkha fertiliser plant. Other credits are not tied to a particular scheme but take the form of programme lending (as with Tunisia and Pakistan) and/or indirect balance-of-payments support in these bilateral arrangements. Moreover, the amount of assistance reported as unallocated (Table 7.6) has been greater than the level of allocated funds in every year from 1973 through 1975; only in 1976 did there appear to be a reversal in this trend. The unallocated funds would seem to indicate lending on more of an ad hoc basis and of the programme-support type. As mentioned earlier, the government of Qatar generally keeps a relatively low profile in publicity on the country's activities; this carries over into the quiet nature of its economic assistance where substantial aid is recorded simply as 'unallocated'.

A second feature of Qatari country-to-country assistance is that the Arab states are the major recipients. Of the nations listed in Table 7.6, only 10 are non-Arab and, more signifi-

Table 7.6: Qatari Bilateral Aid, 1973-6 (in $ millions)

Allocated to:	1973	1974	1975	1976	Total
Bahrain	—	—	0.2	0.4	0.6
Bangladesh	—	1.5	—	—	1.5
Cape Verde Island	—	—	—	0.5	0.5
Egypt	—	5.0	65.2	75.6	145.8
Gabon	—	—	—	1.5	1.5
Gambia	—	—	—	1.1	1.1
Guinea	—	—	4.0	—	4.0
Guinea Bissau	—	1.0	—	—	1.0
Jordan	—	18.5	—	13.9	32.4
Lebanon	—	2.0	0.9	—	2.9
Mali	—	1.5	3.0	2.0	6.5
Mauritania	—	12.0	1.5	1.5	15.0
Morocco	—	8.0	12.0	—	20.0
Oman	2.1	—	12.9	5.5	20.5
Pakistan	—	8.0	3.0	—	11.0
Senegal	—	—	—	1.5	1.5
Somalia	1.0	16.0	2.8	—	19.8
Sudan	—	—	16.0	0.7	16.7
Syria	—	—	0.7	7.1	7.8
Tunisia	—	—	15.0	5.0	20.0
Uganda	—	—	4.0	2.0	6.0
United Arab Emirates	0.5	—	3.5	—	4.0
Yemen (North — Sanaa)	1.1	2.0	—	4.7	7.8
Yemen (South — Aden)	—	—	3.0	2.0	5.0
TOTAL	4.7	75.5	147.7	125.0	352.9
Unallocated[a]	87.9	92.9	155.0	19.0	354.8
Total of allocated and unallocated aid	92.6	168.4	302.7	144.0	707.7

[a] All unallocated funds were directed to Arab countries.

Source: Compiled from Organisation for Economic Cooperation and
Development statistics.

cantly, the so-called 'unallocated' aid (extended but not
identified as to recipient) went to Arab League members.
Non-Arab countries received 7 per cent, 4.6 per cent, and 6
per cent of total Qatari bilateral aid in 1974, 1975, and

1976, respectively.

Qatari aid also flows through collective channels via the various multinational funds to which it contributes and in which it participates financially. Some $ 135 million has been paid in by Qatar to five funds; a sixth, the Arab Authority for Agricultural Investment and Development (AAAID) is still in the formative stages and the degree of Qatari participation has not been finalised. As of the beginning of 1978, Qatar had extended about five times as much aid bilaterally as multilaterally. Nonetheless, by contributing to the collective funds and agencies, Qatari aid takes on a more diversified, less regional, and truly global aspect. A total of 71 countries from Afghanistan to Zambia have received credits from aid bodies to which Qatar has subscribed capital. Moreover, in line with international and regional commitments, Qatar maintains an economic boycott of Rhodesia, South Africa, and Israel.

Qatar is one of 21 Arab League members participating in the Arab Fund for Economic and Social Development (AFESD), the oldest regional collective assistance body, which extends credits to specific projects in Arab nations only.[18]

Qatar's interest in Africa is considerable and takes several forms of multilateral assistance. As one of the 18 subscribers to the Arab Bank for Economic Development in Africa (ABEDA), Qatar has paid in $ 40 million of that body's $ 392.25 million capital. Based in Khartoum, the ABEDA had extended loans of $ 209.6 million as at the end of 1977 to 21 non-Arab African states; the Bank has a potential lending capability of close to $ 1 billion. ABEDA administers the Special Arab Fund for Africa (SAFA) set up in 1976 by the seven oil-exporting Arab countries for emergency balance-of-payments support to African nations in an attempt to relieve the hardships caused to petroleum importers by the 1973 oil price hikes.[19] African countries in oil-caused balance-of-payments straits also have been able to apply for assistance to the OPEC Special Fund, financed by the 13 members (including Qatar) of that organisation. The Vienna-based Special Fund as a $ 1.6 billion capital, to which Qatar contributed $ 36 million. The 57 recipients of OPEC Special Fund assistance (at December 1977) included countries in Africa, Central and South America, the Middle and Far East, ranging from Afghanistan, Burma, Chad and Guatemala to Jamaica, Nepal, Thailand, Mali and the Philippines.

Four other aid bodies have Qatari participation: the Arab Authority for Agricultural Investment and Development (AAAID), the Islamic Development Bank (IDB), the International Monetary Fund's Special Oil Facility, and the Gulf Organisation for Development in Egypt (GODE). The newest of the regional assistance agencies is the AAAID, created at the close of 1977, and geared to the improvement of agriculture throughout the Arab world. Still without headquarters and in the formative stages, the AAAID has an authorised capital of just over $500 million and will first emphasise development of Sudan's agricultural sector.

The Islamic Development Bank's function in the main is equity lending to both public and private projects in Islamic countries or nations having Moslems in their populations. The $2.4 billion capital was paid in by 33 members; Qatar's share was $25 million.[20] Some 14 states had received loans totalling $134.5 million by the end of 1977.

Sensitive to its position as an oil-producing capital-surplus country, Qatar has contributed more than $237 million to the IMF Special Oil Facility (in 1975 and 1977) to assist nations with persistent balance-of-payments deficits attributed to costly oil imports. Finally, Qatar joined with Saudi Arabia, Kuwait, and the United Arab Emirates to form GODE in 1976, a body to handle some $2 billion in aid for stabilising the Egyptian economy.[21] Not only does Egypt hold a significant place in world oil transport and transit, but that country's orderly development and political well-being have very real ramifications for the entire Middle East given its population and strategic location.

Investment Opportunities

Given the rising purchasing power and capital surplus of the Gulf states as a whole, a concomitant interest has evolved in the industrialised nations. There is, of course, the ongoing search for markets by the so-called advanced economies. This search has received further impetus from the growing oil import bills confronting most of the OECD member states; quite simply the balances of trade and payments go hand in hand. Since 1973 there has been a considerable amount written about the 'new international economic order'; raw material producing countries (OPEC) have assumed greater financial power worldwide with a discernible shift of wealth from the industrialised economies to this sector of the developing bloc. With the passing years of this decade, it has become clear that the global trade, monetary and financial patterns are not so much those of confrontation but of interdependence. The need for technology, capital goods, and services in OPEC economies can act to balance the industrialised nations' energy requirements.

In comparison to other capital-surplus Gulf states, Qatar's approach to development is slow and deliberate. Each project is evaluated in terms of its overall contribution to Qatari society. At this stage in the country's development, the opportunities for foreign participation are diverse and range from feasibility studies to construction activities. The current areas of emphasis are those projects contributing to establishment and expansion in social services and physical infrastructure, in addition to some key industrial projects. The strategy for industrialisation is to give priority to those projects which are capital and energy intensive. With the discovery of large quantities of unassociated gas throughout the country,[22] the government of Qatar is encouraging industries which utilise this abundant energy source.

The social services sector has grown rapidly in recent years and will continue to do so in the future. The expansion of the health service, housing, and the educational system, although dependent upon government planning and subsidies, usually relies on foreign technology, management skills, and materials. The situation is similar in the building up of a physical infrastructure. To accommodate the growing needs of a rapidly developing country such as Qatar, the strains placed on the evolving physical infrastructure must be removed. A delay at Doha port for materials arriving for a job site can set back project completion dates which in turn cause a rippling effect throughout the nation for other goals and targets. For this reason, the government of Qatar is pushing and is achieving a modern system of telecommunication and transportation. Referring to the earlier chapter on public finance and budgetary policy, it will be noted that the 1978 budget was reduced by 22 per cent from the level of the previous year's development expenditures. One reason for this reduction is that many projects relating to services, infrastructure, industry, and agriculture have been implemented already with past years' funds utilised. Thus, as time passes, Qatar will complete a modern network of social services, such as hospitals and schools, and establish efficient infrastructure. New projects, repairs, and improvements continue, yet at a slower pace compared to the earlier years of the initial boom.

The area which can only grow as Qatar grows is the services sector. Activities related to insurance, finance and the like will be in steady demand. Opportunities for investment in Qatar will shift therefore, as in any developing country experiencing dynamic transformations.

Although the pace of development in Qatar is somewhat slower than in some of its neighbouring Gulf states, sufficient opportunities exist for the foreign businessman and investor. In addition to a growing economy which is a stimulus for foreign investments and exports, there are various government incentives which encourage participation and cooperation and promote the introduction of that foreign technology deemed appropriate with development goals. This is true of management techniques and imported goods as well. There are, of course, various restrictions to foreign participation. These limitations are twofold resulting both from the nature of Qatar's economy and from government stipulations.

Qatar is subject to certain bottlenecks which are the result of its rapid development. Due to the fact that the rate of development is kept in check by the government, these bottlenecks are controlled somewhat more effectively than in most oil-based economies. Still, problems exist and non-Qataris seeking to undertake trade and financial activities in the state should be aware of them. With the physical infrastructure evolving into a modern system, various deficiencies remain in the structure. The most obvious are the waiting time at Doha port and the housing shortage resulting from the influx of expatriates. And while these problems are being reduced, there naturally exists a time lag between implementation and completion. For this reason, the government chooses to keep the growth of the economy under control so that other sectors can keep pace with the expansion. In

1975 and 1976 the rate of inflation was estimated to be perhaps as high as 25 per cent.[23] The government in Qatar is aware of these hindrances to the business community, and these bottlenecks and problems are taken into account when negotiating ventures and projects.

Government Regulations

A primary goal of government policy is to encourage the establishment of small-scale, joint-venture private manufacturing enterprises which will be suitable for markets in the area and will accommodate local manpower resources. Generally, it is the preferred procedure for Qatari nationals to hold a minimum of 51 per cent of the equity in such companies while import trading and acting as importers' agents and contractors are exclusively in the hands of Qatari nationals. If a company desires exemption from the 51 per cent ownership rules, it must apply for a Decree of Exception from the Ministry of Finance. Should the company be engaged in a joint venture with the government, the Decree would be easy to obtain as it would in the case of an organisation contracted for a specific project in Qatar. The difficulty arises for the firm which wishes to establish a branch in Qatar and seeks the exception. If it can be shown that the company does not compete with established Qatari companies, it may be possible to receive such a decree from the Minister of Finance.

In addition to requiring 51 per cent ownership by Qatari nationals, only Qataris are allowed to own land in the country, a common practice in the Gulf states. The foreign company is provided with a long-term lease or, in the case of a company with a Qatari holding of at least 51 per cent ownership of the capital, the Qatari partner can make arrangements for the site required. Companies which have been given a Decree of Exception depend on leasing property through the private sector and are subject therefore to rent costs.

These restrictions on ownership of capital and land are not uncommon to developing countries and do not appear to have restricted foreign participation. Problems are more apt to be caused from government regulations affecting foreign businessmen that Qatar may be slow to implement due to the newness of an active private sector. For example, it was not until the first part of 1976 that a system of licensing contractors, both foreign and local, was instituted. No system exists at present for the registration of consulting engineers and architects. These firms must state experience and qualification to the various government agencies which they deal directly with. Perhaps the relative compactness and small size of the country have kept the lack of formal devices from having any major negative impact on business ventures in Qatar.

All companies, with the exception of those firms under contract with the government, are subject to income tax. The tax rates listed in Table 7.7 are based on annual profits after deducting general operating expenses. It is possible, if rare, to obtain a waiver for the payment of income tax which must be given directly by the Ruler. Companies are subject only to payment of income taxes as no other taxes exist; thus, for the government to forego income tax from a firm is not common.

Government Incentives

To encourage diversification and the expansion of Qatar's industrial base, the government offers several incentives to companies, both Qatari and foreign, establishing industrial projects in Qatar. Firms which can take advantage of the government-sponsored incentives and loans are those defined as operating in the public interest and not duplicating already established industries. The definition is broad, making for diversity in those companies eligible.

In July 1976, the Industrial Development Technical Center published a report which provided details of the government incentives for industry.

These include:

1. Industrial sites for industrial projects provided at nominal charges.

2. Exemption from income tax for the first five years of operation.

3. Exemption of customs duties on the importation of equipment and material.

4. Utilities, gas, and electricity, in addition to water, are supplied to industrial sites at cost.

Table 7.7: Tax Rates in Qatar

	%
First QR 70,000	Tax free
QR 70,000 – 250,000	5
QR 250,000 – 500,000	10
QR 500,000 – 750,000	15
QR 750,000 – 1,000,000	20
QR 1,000,000 – 1,500,000	25
QR 1,500,000 – 2,000,000	30
QR 2,000,000 – 3,000,000	35
QR 3,000,000 – 4,000,000	40
QR 4,000,000 – 5,000,000	45
QR 5,000,000 +	50

Source: N.A. Shilling, *Doing Business in Saudi Arabia and the Arab Gulf States*
(Inter-Crescent Publishing and Information Corporation, New York, 1975), p. 264.

5. Roads are constructed at government expense to link sites to the main roads.

6. Financial assistance is provided either through guarantees of foreign loans or by arranging for local financing.

7. Protection of the industry may be provided through an increase in tariffs on imports which compete with the industry's product.

8. The government agrees to purchase the output of the company in its entirety or a percentage thereof.

These incentives make it obvious that the government is more than willing to encourage industrial projects undertaken by either Qataris or foreigners. In addition, the incentives attempt to overcome some of the problems Qatar is experiencing as a result of its rapid economic growth. For instance, land prices have skyrocketed around Doha; therefore, the government offers low-cost industrial sites. Furthermore, the problems caused by current development of infrastructure could result in added expenses to project costs. These are reduced by the government providing services, water, and roads to industrial sites free of charge or at government cost. The incentives would appear to be an attempt by the government to equalise investment opportunities between the developed nations and a developing country such as Qatar.

Marketing

The demand for products is on the rise as the economy grows and the expatriate population increases. As mentioned earlier, the Qatari government requires that commercial agencies be entirely Qatari owned. The selection of a local agent by a foreign firm thus can be crucial to the successful marketing of a product. Furthermore, a complex advertising system does not exist in Qatar as it does in Western economies; hence, the agent is relied upon as a critical factor in the promotion of a product. In addition, the agent often serves as a lending organisation by providing credit to customers as banks in Qatar usually do not fund consumer purchases.

Imports into Qatar are growing each year with Japan and the United Kingdom competing as the main exporters to Qatar. In 1976 Japan again captured the largest single share of the Qatar market accountable for 29 per cent of the nation's imports. For that same year, the United Kingdom and the United States accounted for 17 per cent and 8 per cent, respectively. Likewise, West Germany's portion of Qatar's imports was reduced, to 8 per cent. In 1976 France moved up to rank fifth as an exporter to Qatar, having 4 per cent of the market. As described earlier, imports into Qatar consists mainly of machinery, manufactured goods, transport equipment, and building materials.

Qatar has a policy of free trade with import duties set at a low rate, usually around 2.5 per cent. The exceptions rest upon the infant industry argument where duties are raised to protect newly established industries. This protection is used to attract development of new industrial projects for foreign and Qatari businessmen. In addition to the low import duties which are in effect, Qatar has no exchange restrictions on international payments.

When bidding on a construction project or a contract for large purchases of capital equipment it is usually necessary to have a local representative or contracting agency since most tenders are conducted through local agents or on an invitational basis only. Moreover, when tenders are requested, only a short time is given between the announcement and the closing date for bidding. When a foreign company obtains a contract with the government, it is not required to form a partnership with a Qatari national, although they must hire a local agent or agency which is a hundred per cent Qatari owned. The Qatari agent then serves as a liaison between the government and the foreign company providing all of the necessary documents and permits.

When tendering, a company should be aware that the Qatari government considers price the key determinant in offering a contract. Usually no price escalation or cost-plus clauses are considered.

Expatriate Life

As a result of the small population base in Qatar, the country encounters a serious labour shortage. Unskilled workers immigrate from Pakistan, Iran, and India under contracts which specify rate of pay, termination date, employer, and type of work required. It has been estimated that approximately 50 per cent of the population is made up of non-Qataris.[24] The import of labour is relatively easy, requiring only a residence permit from the Immigration Department and a work permit from the Ministry of Labour. These documents are readily obtained as the government recognises the necessity of expatriate workers.

There are of course cultural differences that will require some adjustment in life-styles for expatriates. Liquor is prohibited in Qatar except to the non-Qataris and is very expensive. Outside entertainment, such as movies, private clubs, and restaurants, is limited and expensive. In fact, everything in Qatar is relatively expensive, although inflation is lower than in neighbouring Gulf countries. The shortages caused by the influx of foreigners make it impossible for the country to keep up with the rising demand for certain commodities and services. One such example is the Doha English-speaking school made available for expatriate children. Although the fees are QR 1,300 ($335) per term with three terms per year, the school is full and has a long waiting list. The government has agreed to expand this facility which was designed to accommodate 350 students from the age of five to eleven. For children over eleven, English-speaking schools do not exist and other arrangements must be made for further education.

In conclusion, Qatar as a capital-surplus developing country is growing and utilising technologies from around the world. At present, and for some time in the future, Qatar will rely heavily upon foreign management, development, and maintenance of this technology. In addition, the demand for foreign goods is constantly rising as is the need for a thriving service sector. All this creates a variety of opportunities for the foreign company or businessman. As in any developing country, there are the bottlenecks caused by an infrastructure still being formed. To offset such difficulties, the Qatari government has established incentives to encourage and promote foreign investment.

As a result of the small population base in Qatar, the market for any goods and services will be limited. For this reason the government has encouraged regional cooperation better to utilise the benefits of larger-scale industrialisation.

All in all, it appears that the government is aware of the limitations of industrialisation in Qatar at this stage of its development and because of the constraints of population size and a limited base. As a result, development projects are considered carefully to assure they are consistent with the needs of the country.

Prospects for the Future and Conclusions

The importance of international trade to the economy of Qatar is self-evident. The dominance of one sector (oil) of the economy indeed has been a blessing in the generation of revenues while at the same time it has placed certain obstacles in the country's future. While diversification, industrialisation, and general development continue as the stated objectives of the country, the question remains as to the most effective means of achieving these goals. It is possible that greater regional cooperation will be one of the more attractive paths in the future.

Examination of import trends has revealed overwhelming rates of growth since the early 1970s. Should these trends continue unabated (which is not likely), the present favourable balance-of-trade position of Qatar may be eroded, eventually presenting, as in pre-oil times, a serious balance-of-payments problem. The petroleum sector continues to be the only true means of generating sizeable foreign exchange, and the efforts of the government to diversify the economic base will probably not show marked results for some years to come.

Qatar is similar to other rapidly developing Gulf states in that its present and future development is pegged to oil and in that the distortions in its economy arise from the problems which relatively massive and sudden oil wealth bestows upon a country lacking many of the essential ingredients of development at this stage. Qatar is unique however in its moderate approach to economic development. It does not hold the ambition of fashioning Doha into the financial capital of the Gulf region, as neighbouring countries like Kuwait and Bahrain hope to do for their capital cities. Neither is it the policy of the government to establish projects for reasons of prestige. Consequently, when it is found that a prospective project has lost its economic feasibility, the authorities concerned do not hesitate to revise estimates downwards or to scrap it entirely.

The low-key development profile which Qatar has adopted has both its costs and its benefits. It has delayed, for example, international economic recognition of this small but highly development-oriented member of the Gulf community. Lacking sufficient recognition, the country might have found itself at a disadvantage when competing for scarce inputs in development, such as skilled manpower, technology and entrepreneurship, especially from American sources. It could be pointed out, however, that this may have been offset by the esteem in which Britain holds Qatar owing to the colonial relationship that existed between the two countries.

A rational approach towards project appraisal and implementation and the resulting relatively moderate pace of economic development have enabled the government to avoid unnecessary mistakes and by so doing, come out with economically feasible and socially acceptable development programmes without relying on a comprehensive development plan. Also, the relative price stability which the country has been able to achieve has made Qatar an attractive place for international firms to do business. Above all, the development strategy has enabled the government to give sufficient attention to those programmes which actually raise the social and economic well-being of the population.

The analysis of the structural development of the economy indicates remarkable progress; economic development in Qatar has taken place without a comprehensive development plan. An achievement worth emphasising is the success of the agricultural sector. Within a short span of time and despite the physical limitations posed by scarcity of water and cultivable land, Qatar has emerged from a position of shortage to one of self-sufficiency in vegetables and fruits. The fishing industry also has shown impressive results.

The diversification policy of the country however has been adversely affected by the inability of the industrial sector to respond fast enough and on a sufficient scale to the incentives and the development measures of the government. The success which has been recorded in this sector lies in the area of oil-based projects. Qatar sustained a setback when fire destroyed an important plant in the petrochemical sector. Qatar is not alone in the struggle to reduce its dependence on crude oil exports. The country may even be considered better off than some of its neighbours to the extent that it has substantial deposits of non-associated gas upon which a viable petrochemical industry can be erected to replace oil in the future.

The success of such a venture will rely heavily on close international cooperation. Yet, the dependence of Qatar on trade relations with the rest of the world is not surprising, given the limited resource base of the country today and its past history. Knowledge of this has given rise to a liberal international trade policy. Without doubt the liberal trade policy has been accentuated further by the balance-of-payments surpluses which have become a normal feature of the Qatari economy. So far, imports have not been able to rise fast enough to absorb completely the foreign exchange from oil exports, despite the liberal policy and the increased development activities. In the future, however, should the imports sector continue its expansionary trend while exports at best remain constant, the size of the surpluses will gradually diminish. Unless non-traditional exports can be expanded to offset the gradual depletion of the finite oil reserves, deficits in the balance of payments could occur in the future, albeit in the distant future. For this reason, the attempt to reorientate the economy towards non-associated gas and petrochemical industries is a move in the right direction. Indeed, it appears that this area of diversification holds the major hope for sustaining the balance of payments position of the country and for ensuring an acceptable rate of growth of the economy over coming decades.

The liberal trade policy means, however, that government revenue from taxes on imports and other types of custom duties is necessarily low. In addition, other non-oil sources of revenue still constitute an insignificant part of government revenue. The economic effects of oil depletion in the long run thus include falling government revenues and, unless diversification and other income-generating activities have evolved, social services could diminish. A case could be made for the 'learning by doing' form of taxation which would make it easier for the government to institute an alternative system of taxation in the future when the oil is exhausted.

As is found in studies concerning oil-generated surplus-funds countries, the main constraint on Qatari development is manpower. An adequate supply of labour from both qualitative and quantitative frames of reference is hard to come by basically because the nation cannot afford to allow an unrestricted influx of foreign labour owing to the socio-economic implications which domestic-foreign labour or population imbalance would generate. The future absorptive capacity of Qatar will be determined largely by the extent of success which the government achieves in balancing two opposing factors: the demand for foreign manpower and the disutilities which are the by-products of an excessive inflow of foreign labour.

In the final analysis, it appears that the ability of Qatar and, indeed, other Gulf states to cooperate in both economic and social development programmes will test their desire for an ongoing, flourishing economy and ultimately perhaps, even economic survival. The theoretical benefits of and the problems of economic cooperation are many and varied. Of special importance are the production and marketing of petrochemical products. Considering that the total world market for petrochemical products is fixed at a point in time (or from the dynamic point of view, it can grow only by a fixed rate), coordination in this field by the Gulf states is a precondition for preventing a glut in the world petrochemical market, unless the advanced market economies are prepared to reduce production which they are unlikely to do. Further, a functioning level of regional cooperation would benefit the development and transfer of technological know-how in the petrochemical industry and other activities.

This analysis has attempted to identify the growth points of the Qatari economy, examine the sectoral structure, policy goals, problems of economic development and limited absorptive capacity. Suggestions for alleviating some of the development constraints and for ensuring the future prosperity of the country have been offered. This concern for coming generations underpins the Qatari attitude to development and, despite the temptation offered by a capital surplus to undertake massive projects and transplant technology and other economic factors irrespective of cost, the restraint and moderation of Qatar offers an example which might well be emulated by other oil economies.

Notes

1. While the paucity of recent trade data prevents an up-to-date discussion of disaggregated import trends, the years 1969 to 1971 utilised in these calculations are the most current available; they are reasonable indicators of what is believed to be the general situation over the last decade (i.e., a shift from consumption to capital goods imports). See State of Qatar, Ministry of Economics and Commerce, *The Economic Supply for 1971* (Doha, 1976), pp. 72-4.

2. The source is the United States Department of Commerce as quoted in the *Bulletin of the American-Arab Association*, vol. VIII, no. 8, October 1977, p. 1.

3. Total crude oil exports declined from 570.3 thousand barrels per day in 1973 to 511.2 thousand barrels per day in 1974.

4. M.O. Clement *et al.*, *Theoretical Issues in International Economics* (Houghton Mifflin Company, Boston, Mass., 1976), p. 178.

5. Note that the original cost of production to a non-member may be lower than the cost of production of the union partner, but with the existence of the tariff, the partners' goods will be cheaper in the union market than those of the non-members.

6. This issue and the following considerations have a direct bearing on successful adoption of proposed integration in the Gulf region.

7. M.O. Clement *et al.*, *Theoretical Issues*, p. 186.

8. Jacob Viner, *The Customs Union Issue* (Carnegie Endowment for International Peace, New York, 1950), p. 35.

9. G. Makower and G. Morton, 'A Contribution Towards a Theory of Customs Unions', *Economic Journal*, March 1953.

10. M.O. Clement *et al.*, *Theoretical Issues*, p. 199. E.A.G. Robinson (ed.), *Economic Consequences of the Size of Nations*, proceedings of a conference held

by the International Economic Association (Macmillan and Co., London, 1960) offers a number of chapters which outline the special problems of the small economy as, for example, those by T. Scitovsky, 'International Trade and Economic Integration as a Means of Overcoming the Disadvantages of a Small Nation', and G. Marcy, 'How Far Can Foreign Trade and Customs Agreements Confer upon Small Nations the Advantages of Large Nations?'

11. Bela Belassa, *The Theory of Economic Integration* (Richard D. Irwin, Inc., Homewood, Illinois, 1961), pp. 164-5.

12. M.O. Clement *et al., Theoretical Issues*, p. 199.

13. Ministry of Economics and Commerce, Abu Dhabi, 'The Gulf Common Market' (July 1973, in Arabic), advocates a common market between Kuwait, Qatar, Bahrain, the United Arab Emirates, and Oman.

14. MEED, 22 April 1977, p. 18. An interview with the president of the Arab Monetary Fund, *New York Times*, 20 June 1978, offers details on the objectives and resources of the Fund.

15. The many national and collective assistance agencies are briefly reviewed and assessed in Ragaei El Mallakh and Mihssen Kadhim, 'Arab Institutionalized Development Aid: An Evaluation', *The Middle East Journal*, autumn 1976, pp. 471-84. Some of the more theoretical and general aspects of the capital surplus and deficit Arab states and the issue of absorptive capacity which bears on the *raison d'être* for development lending by and to Middle Eastern nations are outlined in Ragaei El Mallakh and Mihssen Kadhim, 'Capital Surpluses and Deficits in the Arab Middle East: A Regional Perspective', *International Journal of Middle East Studies*, April 1977, pp. 183-93, and 'Absorptive Capacity, Surplus Funds, and Regional Capital Mobility in the Middle East', *Rivista Internazionale de Scienze Economiche e Commerciali*, vol. 24, no. 4 (1977), pp. 308-25.

16. Members of the OECD are: Australia, Austria, Belgium, Canada, Denmark, Finland, France, the Federal Republic of Germany, Greece, Iceland, Ireland, Italy, Japan, Luxembourg, the Netherlands, New Zealand, Norway, Portugal, Spain, Sweden, Switzerland, Turkey, the United Kingdom, and the United States. The aid-extending OECD members form the Development Assistance Committee and include: Austria, Belgium, Canada, Denmark, France, Germany, Italy, Japan, the Netherlands, Norway, Portugal, the United Kingdom, and the United States.

17. John D. Law, *Arab Aid: Who Gets It, For What, and How* (Chase World Information Corporation, New York, 1978), p. xiv.

18. The authorised potential capital of the AFESD (headquartered in Kuwait), utilising its borrowing provisions and the like, is $ 4.14 billion.

19. Of the SAFA's $ 360 million capital, all but about $ 10 million had been loaned by the beginning of 1978. Qatar contributed $ 20 million to this Fund. In addition, at the March 1977 Afro-Arab Summit Conference held in Cairo, Qatar pledged $ 118 million toward long-term economic development in Africa. MEES, 6 March 1978, p. 7.

20. The IDB has headquarters in Jeddah. The recipients of loans have varied from Algeria and other Arab countries to Bangladesh, Cameroon, Guinea, Niger, Pakistan, Senegal, and Turkey.

21. Qatar holds 15 per cent of the Organisation.

22. See the chapter on the development of the oil industry in Qatar.

23. N.A. Shilling, *Doing Business in Saudi Arabia and the Arab Gulf States, 1977 Supplement* (Inter-Crescent Publishing and Information Corporation, New York, 1977), p. 71.

24. MEED, special report, April 1977, p. 23.

SELECT BIBLIOGRAPHY

Books

Abolfathi, Farid *et al., The OPEC Market to 1985* (D.C. Heath and Co., Lexington, Mass., 1977)

Adler, John H., *Absorptive Capacity: The Concept and its Determinants* (The Brookings Institution, Washington, DC, 1965)

Anthony, John Duke, *Arab States of the Lower Gulf: People, Politics, Petroleum* (The Middle East Institute, Washington, DC, 1975)

Belassa, Bela, *The Theory of Economic Integration* (Richard D. Irwin, Inc., Homewood, Illinois, 1961)

Bibby, Geoffrey, *Looking for Dilmun* (Alfred A. Knopf, New York, 1969)

Clement, M.O. *et al., Theoretical Issues in International Economics* (Houghton-Mifflin Company, Boston, Mass., 1976)

Edo, Michael E., *Currency Arrangements and Banking Legislation in the Arabian Peninsula* (International Monetary Fund, Middle Eastern Department, Washington, DC, 1974)

El Mallakh, Ragaei, *Economic Development and Regional Cooperation: Kuwait* (University of Chicago Press, Chicago, 1968)

El Mallakh, Ragaei (ed.), *Energy Options and Conservation*, proceedings of the Fourth International Energy Conference (International Research Center for Energy and Economic Development, University of Colorado, Boulder, Colorado, forthcoming, 1978)

El Mallakh, R. and Carl McGuire (eds.), *Energy and Development*, proceedings of the First International Conference (International Research Center for Energy and Economic Development, University of Colorado, Boulder, Colorado, 1974)

El Mallakh, R. and Carl McGuire (eds.), *US and World Energy Resources: Prospects and Priorities*, proceedings of the Third International Energy Conference (International Research Center for Energy and Economic Development,

University Of Colorado, Boulder, Colorado, 1977)

El Mallakh, R., Mihssen Kadhim and Barry Poulson, *Capital Investment in the Middle East* (Praeger Publishers, New York, 1977)

El-Rayyes, R.N., *The Oasis Rivalry and Oil: The Problems of the Arabian Gulf, Between 1968-1972* (El Harar, Beirut, 1973) (in Arabic)

Fallon, Nicholas, *Middle East Oil Money and its Future Expenditure* (Graham Trotman Ltd, London, 1975)

Frank, Helmut J., *Crude Oil Prices in the Middle East* (Frederick A. Praeger, New York, 1966)

Higgins, Benjamin, *Economic Development: Problems, Principles and Policies* (revised edition) (W.W. Norton and Co., Inc., New York, 1968)

Jacoby, Neil H., *Multinational Oil* (Macmillan Publishing Company, Inc., New York, 1974)

Kindleberger, Charles D., *Economic Development* (second edition) (McGraw-Hill Book Company, New York, 1965)

Kubbah, Abdul Amir, *OPEC Past and Present* (Petro-Economic Research Centre, Vienna, September 1974)

Law, John D., *Arab Aid: Who Gets It, For What, and How* (Chase World Information Corporation, New York, 1978)

Meier, G.M., *Leading Issues in Economic Development* (third edition) (Oxford University Press, New York, 1977)

Mercier, Claude, *The Petrochemical Industry and the Possibilities of its Establishment in the Developing Countries* (Editions Techniq, Paris, 1966)

Monroe, Elizabeth (ed.), *The Changing Balance of Power in the Persian Gulf* (The American Universities Field Staff, Inc., New York, 1972)

Musrey, Alfred G., *An Arab Common Market* (Frederick A. Praeger, New York, 1969)

Park, Yoon S., *The World Oil Economy in the 1970s* (Westview Press, Boulder, Colorado, 1976)

Robinson, E.A.G. (ed.), *Economic Consequences of the Size of Nations* (Macmillan and Company, London, 1960)

Sadik, Muhammed T. and William P. Snavely, *Bahrain, Qatar, and the United Arab Emirates* (D.C. Heath and Company, Lexington, Mass., 1972)

Sharbiny, Naiem A. and Mark A. Tessler (eds.), *Arab Oil* (Praeger Publishers, New York, 1976)

Shilling, N.A., *Doing Business in Saudi Arabia and the Arab Gulf States* (IPIC, New York, 1975)

Shilling, N.A., *Doing Business in Saudi Arabia and the Arab Gulf States, 1977 Supplement* (IPIC, New York, 1977)

Singer, Hans W., *International Development: Growth and Change* (McGraw-Hill Publishers, New York, 1964)

Snider, Delbert A., *Introduction in International Economics* (fourth edition) (Richard D. Irwin, Inc., Homewood, Illinois, 1967)

Tobin, James and William D. Nordhaus, *Economic Growth* (National Bureau of Economic Research (NBER), NBER Colloquia Series on Economic Research: Retrospect and Prospect, no. V, New York, 1972)

Viner, Jacob, *The Customs Union Issue* (Carnegie Endowment of International Peace, New York, 1950)

Official Sources, Government, and International Agencies

Abu Dhabi, Ministry of Economics and Commerce, *Economics and Commerce* (Abu Dhabi, July 1973) (in Arabic)

Abu Dhabi, Ministry of Economics and Commerce, *The Gulf Common Market* (Abu Dhabi, July 1973) (in Arabic)

International Monetary Fund (IMF), *International Financial Statistics* (IMF, Washington, DC)

International Monetary Fund, *IMF Survey* (IMF, Washington, DC)

Permanent Mission of the State of Qatar to the United Nations, *The Era of Reform* (United Nations, New York, 1973)

Qatar, Customs Department, *Yearly Bulletin of Imports, Exports, and Transit* (Doha, Qatar, 1977)

Qatar, Department of Petroleum Affairs, *Oil Industry in Qatar 1971* (Qatar National Printing, Doha, Qatar, n.d.)

Qatar, Department of Training and Career Development, *Manpower in Qatar* (Qatar National Printing, Doha, Qatar, March 1974)

Qatar, Ministry of Economics and Commerce, *An Analysis of the Commodity Imports to Qatar* (Dar El-Ulum Establishment, Doha, Qatar, 1972)

Qatar, Ministry of Economics and Commerce, *The Economic Supply for 1971* (Qatar National Printing, Doha, Qatar, 1976)

Qatar, Ministry of Information, *Qatar into the Seventies* (Ali bin Ali Printing Press, Doha, Qatar, 1973)

United Nations Economic and Social Office in Beirut (UNESOB), *United Nations Inter-disciplinary Reconnaissance Mission to Qatar* (United Nations, Beirut, July 1972)

United Nations Industrial Development Organization, *An Industrial Survey of Qatar* (United Nations, New York, April 1972)

United States Department of State, *Medium Term Ability of Oil-Producing Countries to Absorb Real Goods and Services* (CACI, Inc., Arlington, Va., March 1976)

Articles, Journals, Bulletins, Papers

Annual Statistical Bulletin (OPEC) (Information Department, Organization of Petroleum Exporting Countries, Vienna, annual)

Bulletin of the American-Arab Association (American-Arab Association for Commerce and Industry, Inc., New York, monthly)

Economist Intelligence Unit, *Middle East Annual Review* (Middle East Review Co. Ltd, Essex, England, annual)

El Mallakh, Ragaei and Mihssen Kadhim, 'Absorptive Capacity, Surplus Fund and Regional Capital Mobility in the Middle East', *Rivista Internazionale de Scienze Economiche e Commerciali*, vol. 24, no. 4, 1977

El Mallakh, Ragaei and Mihssen Kadhim, 'Arab Institution-alized Development Aid: An Evaluation', *The Middle East Journal*, autumn 1976

El Mallakh, Ragaei and Mihssen Kadhim, 'Capital Surpluses and Deficits in the Arab Middle East: A Regional Perspective', *International Journal of Middle East Studies*, April 1977

Horvat, Branko, 'The Optimum Rate of Investment', *Economic Journal*, December 1958

Makower, G. and G. Morton, 'A Contribution Towards a Theory of Customs Union', *Economic Journal*, March 1953

Middle East Economic Digest (Middle East Economic Digest Ltd, London, weekly)

Middle East Economic Survey (Middle East Petroleum and Economic Publications, Nicosia, Cyprus, weekly)

Middle East Surveys (The Financial Times Ltd, London, November 1976 to August 1977)

Mideast Markets (Chase World Information Corporation, New York, every other week)

OAPEC News Bulletin (Information Department, Organization of Arab Petroleum Exporting Countries, Kuwait, monthly)

Oil & Gas Journal (Petroleum Publishing Company, Tulsa, Oklahoma, weekly)

OPEC Bulletin (Information Department, Organization of Petroleum Exporting Countries, Vienna, weekly)

OPEC Review (Public Relations Department, Organization of Petroleum Exporting Countries, Vienna)

Petroleum Economist (Petroleum Press Bureau Ltd, London, monthly)

Petroleum Intelligence Weekly (Petroleum and Energy Intelligence Weekly, Inc., New York, weekly)

Qatar News (Embassy of the State of Qatar, Washington, DC, six times a year)

The Journal of Energy and Development (International Research Center for Energy and Economic Development, Boulder, Colorado, semi-annual)

Voice of the Arab World/Voice (Morris International Ltd, London)

INDEX

absorptive capacity 28, 57, 59, 73, 76-80, 151; limitations on 72, 78; Qatar's 76-80; the 'human factor' and 78, 80

Abu Dhabi 17, 34, 38, 40, 122

Abu Dhabi Marine Areas (ADMA) 34

Adler, John 77, 78

Advisory Council 17

Afghanistan 157, 158

Africa 76, 145, 157; imports from 142

agriculture 72, 89-93, 168; AAAID formation for improvement of 158; agricultural potential of Oman and Ras al-Khaimah 151; foreign labour domination in 24, 25, 27; government assistance to 57, 89, 96; labour and labour productivity in 24-6, 27, 90; unfavourable prospects for expansion of 75, 89

aid (Qatari) 10, 65, 66, 67, 153-8; as percentage of GNP 154; bilateral 156; recipients of 155; summary of Qatari aid 155; types and amounts of 154-5

Ajman 111

Alaska 44

Algeria 101

Al Khalifa (Khalifa family) 15

Al Thani family 15-16

Al Thani, Sheikh Abdel-Aziz bin Khalifa 38, 151

Al Thani, Sheikh Abdullah 16

Al Thani, Sheikh Jasim 16

Al Thani, Sheikh Khalifa bin Hamed 16, 56, 151

Al Thani, Sheikh Muhammad (bin Thani) 15, 16

aluminium smelting 75

ammonia 43, 80, 83

Arab-African Oil Assistance Fund 154

Arab Authority for Agricultural Investment and Development (AAAID) 157, 158

Arab Bank for Economic Development in Africa (ABEDA) 157

Arab Fund for Economic and Social Development (AFESD) 154, 157

Arab League (League of Arab States) 16, 153, 156, 157

Arab Maritime Petroleum Transport Company 37, 53

Arab Monetary Fund (AMF) 146, 153

Arab Petroleum Investment Corporation 53, 153

Arab Shipbuilding and Repair Yard Company 37, 53

Arab States Broadcasting Union (ASBU) 101

Arab Tankers Company 153

Arabian Sea 9, 18

archaeological expeditions to Qatar 18

Austria 113

Awamir tribe 20

Bahrain 17, 19, 51, 99, 100, 101, 104, 105, 122, 145, 151, 153, 156, 167

balance of payments 123, 144-5, 158, 166, 168

Bangladesh 156

Bani Hajir tribe 20

banking 119, 122-3, 125, 127-31; commercial 122-3, 127-31; development of 119; foreign dominance of and in insurance 24, 123, 126-7; Qatari definition of 138

Bedouins 20, 89

Brazil, imports from 142

British Airports Authority 98

British Petroleum 19, 33, 34

British Power Gas Corporation 82

budgets (Qatari): development budgets 69, 159-60; government control of 68; revenues 58-65; surpluses 58, 59, 60, 61, 63

Bul Hanine field, oil production 33, 35, 38, 40

Bunduq field 34, 40

Burma 158

Bushire 18

177